Liquid Scintillation

SCIENCE AND TECHNOLOGY

D1486318

Academic Press Rapid Manuscript Reproduction

Proceedings of the International Conference on
Liquid Scintillation: Science and Technology
held at the Banff Centre in Alberta, Canada
on June 14–17, 1976

Liquid
Scintillation

SCIENCE AND TECHNOLOGY

Edited by

A. A. Noujaim
C. Ediss
L. I. Weibe

Faculty of Pharmacy and Pharmaceutical Sciences
University of Alberta
Edmonton, Alberta
Canada

Academic Press, Inc.   New York   San Francisco   London   1976

A Subsidiary of Harcourt Brace Jovanovich, Publishers

ACADEMIC PRESS, INC.
111 Fifth Avenue, New York, New York 10003

United Kingdom Edition published by
ACADEMIC PRESS, INC. (LONDON) LTD.
24/28 Oval Road, London NW1

LIBRARY OF CONGRESS CATALOG CARD NUMBER: 76–44559

ISBN 0–12–522350–1

PRINTED IN THE UNITED STATES OF AMERICA

Contents

CONTENTS

List of Contributors

Abrams, D.N., Faculty of Pharmacy and Pharmaceutical Sciences, University of Alberta, Edmonton, Alberta, T6G 2H7, Canada

Apelgot, S., Laboratoire Curie de la Fondation Curie, Institut du Radium, 11 Rue P. et M. Curie, 75231 Paris, Cedex 05, France

Bransome, E.D. Jr., Department of Medicine, Medical College of Georgia, Augusta, Georgia 30902, U.S.A.

Buess, E.M., Division of Biological and Medical Research, Argonne National Laboratory, Argonne, Illinois 60439, U.S.A.

Burnham, J., Manager, Health Physics, New Brunswick Power, 527 King Street, Fredericton, New Brunswick, E3B 4X1, Canada

Carter, T.P., Faculty of Medicine, Radioisotope Counting Services, Medical School, University of Birmingham, Engrasson, Birmingham, B15 2TJ, England

Demulyder, F., Vrije Universiteit Brussel, Laboratorium voor Biochemie, Brussels, Belgium

De Rycker, J., Vrije Universiteit Brussel, Laboratorium voor Biochemie, Brussels, Belgium

Duquesne, M., Laboratoire Curie de la Fondation Curie, Institut du Radium, 11 Rue P. et M. Curie, 75231 Paris, Cedex 05, France

Ediss, C., Faculty of Pharmacy and Pharmaceutical Sciences, University of Alberta, Edmonton, Alberta, T6G 2H7, Canada

English, D., W.W. Cross Cancer Institute, Edmonton, Alberta, T6G 1Z2, Canada

Erwin, W., Laboratory of Chemical Biodynamics, Lawrence Berkeley Laboratory, University of California, Berkeley, California 94720, U.S.A.

Ferris, R., Amersham/Searle Corp., 2636 S. Clearbrook Drive, Arlington Heights, Illinois 60005, U.S.A.

Gibson, J.A.B., Environmental and Medical Sciences Division, Atomic Energy Research Establishment, Harwell, Oxon, England

Gordon, B.E., Laboratory of Chemical Biodynamics, Lawrence Berkeley Laboratory, University of California, Berkeley, California 94720, U.S.A.

Horan, T., W.W. Cross Cancer Institute, Edmonton, Alberta T6G 1Z2, Canada

Horrocks, D.L., Scientific Instruments Division, Beckman Instruments Inc., Irvine, California 92713, U.S.A.

Kelly, M.J., Beckman Instruments Inc., Scientific Instruments Division, Irvine, California 92713, U.S.A.

Kisieleski, W.E., Division of Biological and Medical Research, Argonne National Laboratory, Argonne, Illinois 60439, U.S.A.

Kohler, V., Scientific Instruments Division, Beckman Instruments Inc., Irvine, California, 92713, U.S.A.

Laney, B.H., Searle Instrumentation, 2000 Nuclear Drive, Des Plaines, Illinois 60018

Lemmon, R.M., Laboratory of Chemical Biodynamics, Lawrence Berkeley Laboratory, University of California, Berkeley, California 94720, U.S.A.

Long, E., Beckman Instruments Inc., Scientific Instruments Division, Irvine, California 92713, U.S.A.

McDowell, W.J., Chemical Technology Division, Oak Ridge National Laboratory, Oak Ridge, Tennessee 37830, U.S.A.

McPherson, T.A., W.W. Cross Cancer Institute, Edmonton, Alberta, T6G 2H7, Canada

McQuarrie, S.A., Faculty of Pharmacy and Pharmaceutical Sciences, University of Alberta, Edmonton, Alberta T6G 2H7, Canada

Noakes, J.E., Geochronology Laboratory, University of Georgia, Athens, Georgia 30601, U.S.A.

Noujaim, A.A., Faculty of Pharmacy and Pharmaceutical Sciences, University of Alberta, Edmonton, Alberta T6G 2H7, Canada

Painter, K., Dept. of Radiology and Radiation Biology, Colorado State University, and Micromedic Diagnostics Inc., Fort Collins, Colorado 80521, U.S.A.

Peng, C.T., Department of Pharmaceutical Chemistry, School of Pharmacy, University of California, San Francisco, California 94143, U.S.A.

Polic, E., Packard Instrument, Downers Grove, Illinois 60515, U.S.A.

Press, M., Laboratory of Chemical Biodynamics, Lawrence Berkeley Laboratory, University of California, Berkeley, California 94720, U.S.A.

Reid, W., University Hospital, Saskatoon, Saskatchewan, Canada

Roosens, H., Vrije Universiteit Brussel, Laboratorium voor Biochemie, Brussels, Belgium

Ross, H.H., Analytical Chemistry Division, Oak Ridge National Laboratory, Oak Ridge, Tennessee 37830, U.S.A.

Schram, E., Vrije Universiteit Brussel, Laboratorium voor Biochemie, Brussels, Belgium

Stanley, P.E., Department of Clinical Pharmacology, The Queen Elizabeth Hospital, Woodville, South Australia 5011

Wallick, E.I., The Alberta Research Council, 11315 - 87 Ave., Edmonton, Alberta, Canada

Weiss, J.F., University of Tennessee, Comparative Animal Research Laboratory, Oak Ridge, Tennessee 37830, U.S.A.

Wiebe, L.I., Faculty of Pharmacy and Pharmaceutical Sciences, University of Alberta, Edmonton, Alberta, T6G 2H7, Canada

Wigfield, D.C., Department of Chemistry, Carleton University, Ottawa, Ontario, Canada

Preface

This volume contains the proceedings of the International Conference on Liquid Scintillation: Science and Technology held on June 14-17, 1976 at the Banff Centre in Alberta, Canada. A majority of the 70 delegates who attended the conference stayed in the accommodation provided on the Banff campus and enjoyed its beautiful setting in the heart of the Rocky Mountains. Despite the distracting surroundings, the conference program was full and intensive. Excellent plenary lectures were given by Drs. Horrocks, Ross, Gibson, and Stanley. Unfortunately, our fifth plenary lecturer, Mr. H.A. Polach, was unable to attend the conference due to sudden illness. We missed both his presence and his news from the world of low background counting.

Although liquid scintillation counting is a fairly mature technique, research papers were presented at the conference on a wide range of topics including not only old problems such as the counting of heterogenous samples, but also the counting of novel isotopes and bioluminescence. We would like to express our gratitude to Dr. C.T. Peng, Dr. E. Schram, and Dr. P.E. Stanley, who together with members of the organizing committee chaired the research sessions.

The idea of workshops or panel discussions on liquid scintillation counting introduced at the Sydney symposium in 1973 was expanded to two such sessions, one on instrumentation and one on sample preparation. Much of the credit for the lively discussions must be given to the panel members (Instrumentation: B.H. Laney—Chairman, J.E. Noakes, and P.E. Stanley. Sample preparation: E.D. Bransome Jr., R. Ferris, W.E. Kisieleski, K. Painter—Chairman, C.T. Peng, and D.C. Wigfield) who presented the introductory workshop papers that are included in these proceedings. The ensuing discussion periods provided a forum for not only the further consideration of current problems but also informal speculation into the future.

For providing financial assistance the organizing committee would like to sincerely thank Searle Instrumentation, Packard Instrument Inc., Beckman Instruments Inc., and Terochem Laboratories. We are very grateful to both the Province of Alberta and the University of Alberta for also providing financial assistance. We offer our special thanks to Dr. M.J. Huston, Dean of the Faculty of Pharmacy and Pharmaceutical Sciences at the University of Alberta not only for assisting financially but also for allowing the conference to use the faculty secretarial staff. Also, from our faculty we would like to sincerely thank the Associate Dean, Dr. G.E. Myers, for initiating the conference proceedings.

For their guidance in the initial planning stages and their support throughout, we thank the members of the Scientific Advisory Committee (S. Apelgot, J.B. Birks, C.T. Peng, H.H. Ross, E. Schram, and P.E. Stanley).

We are sure that all the conference delegates would like to express their appreciation to the staff and management of the Banff Centre, and to Mr. Bob Sandford of the National Parks interpretive program who gave us an entertaining and enlightening talk about Banff. Also, in a lighter vein we compliment Dr. H.H. Ross who is not only an expert in Cerenkov counting but also exhibited great talent as a chef at our barbecue. This pleasantly informal occasion allowed an opportunity to greet old friends and make new ones.

Finally, we must express our gratitude to those people behind the scenes. Mr. J. Mercer, who helped with the lighting at the conference, and Mr. S. McQuarrie, who not only helped with the audio visual equipment but also has made a great contribution in proofreading and indexing the proceedings. Our secretaries, Bev Johnson and Pearl Metke, also deserve our sincerest thanks for their invaluable help, both during and after the conference.

A.A. Noujaim
C. Ediss
L.I. Wiebe

The Mechanisms of the Liquid Scintillation Process

by

Donald L. Horrocks
Scientific Instruments Division
Beckman Instruments Inc.
Irvine, California 92713

The liquid scintillation process is based upon the conversion of part of the kinetic energy of an ionizing particle (usually from the decay of a radionuclide) into photons. These photons are collected and measured by multiplier phototubes (MPTs) and subsequently, the pulses from the MPTs are summed, sorted, and counted. Initially, the liquid scintillation process was used only as a means of detecting and quantitating the amount of radionuclide present in a sample. However, in recent years, there have been more and more applications involving not only the detection and quantitating, but also the measure of the distribution of the amplitudes of pulses produced by the interaction of the radiations with the liquid scintillators. Since the pulse amplitude can be calibrated with the energy of the radiations, it is possible to measure more than one radionuclide in a sample by selection of the pulse amplitudes emanating from the different radionuclides. Figure 1 shows a pulse height spectrum of a sample of thorium showing the presence of several radionuclides which can be identified by the different pulse amplitudes produced in the liquid scintillator.

In the history of liquid scintillation, it seems as though the number of "users" of liquid scintillators has increased at a tremendous rate. However, as is the case with many analytical methods, many "users" do not totally appreciate the complexity of the liquid scintillation process and as a result, often misinterpret results or attempt experiments which are beyond the capability of the method.

In this paper, it will be attempted to discuss the many mechanisms which comprise the "liquid scintillation process". It is hoped that an understanding of the mechanisms will help the many "users" to obtain more meaningful and accurate data from their experiments.

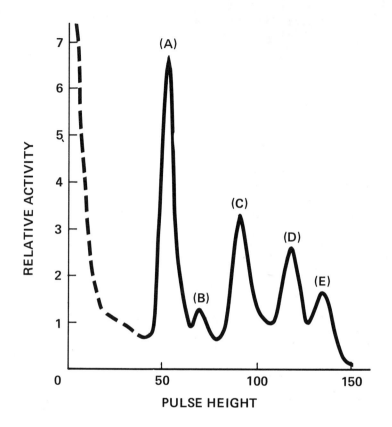

DIFFERENTIAL PULSE HEIGHT SPECTRUM FOR Th^{232}, Th^{228}, AND DAUGHTERS PLUS A SMALL AMOUNT OF Th^{230}.

Figure 1.   Pulse height spectrum of Thorium sample
              dissolved in a liquid scintillation solution.

Interaction of Ionizing Radiation

The energy or distribution of energies of the radiations
from a radionuclide are always the same. They are, with
only a few exceptions, unaltered by the nature of the sample.
However, the response that they produce in a liquid scintillator will depend upon many factors. The first factor to
keep in mind is that only that energy or part of the energy
which is released to the liquid scintillator can contribute
to the response that will be produced. Any energy which is
carried out of the liquid scintillator solution will be
lost. This fact is often overlooked when measuring radionuclides which emit particles with high energies or when
using small volumes of the liquid scintillator. In all
liquid scintillation counting with homogeneous distribution
of the radionuclide, there is always a fraction of radionuclides near the walls of the container. Part of these
radiations can reach the wall before losing all of their
energy. Thus, the response produced will be less than that
produced by a particle of the same energy which releases all
of its energy to the liquid scintillation medium. This is
often called the "wall effect". The wall effect will have
only a slight effect on counting in a wide open counting
channel (accepting all pulse heights), but can markedly alter
the relative counts in narrow counting channels (accepting
only a fraction of the pulse heights).

If the sample is not homogeneously distributed within the
liquid scintillator, the energy released to the scintillation-
producing medium can also be reduced. This effect is most
commonly encountered when the radionuclide is in a second
phase such as deposited on a filter paper, a gel slice, a
precipitate, etc. In these cases, part of the energy is
released in the solid matrix before it reaches the liquid
scintillator system. Thus, the radiations interact as if
they originate with less energy. This effect is often
referred to as "self-absorption". If the radiation energy
is totally released within the second phase, there will be
no response produced within the liquid scintillator.

When the scintillating solvent system is diluted by sample
or secondary solvents, part of the energy may be released in
the non-scintillation producing diluents. For a given
energy, less excited solvent molecules will be produced with
a smaller number of photons being emitted from the liquid
scintillator. Thus, the response produced will be only
equivalent to the response produced by a less energetic

event in the undiluted liquid scintillator. This effect is referred to as "dilution absorption". (As will be explained later, dilution can also effect the energy migration and energy transfer processes.)

Another type of process which leads to decreased excited solvent molecules can be referred to as "annihilation". This effect is experienced when the total energy is removed in one catastrophic event. This effect can be noted in samples which contain heavy atoms (high atomic number atoms) which, when the ionizing particle strikes the heavy atom, absorb all of the kinetic energy without producing excited solvent molecules. Depending upon how far the particle has traveled in the liquid scintillator before it encounters such an "annihilation", the response will be proportionally decreased.

In liquid scintillator solutions, the primary excitations occur in the solvent part of the solution. The final response is directly proportional to the number of excited solvent molecules produced in this initial step. Of course, as already discussed, many processes can inhibit the production of excited solvent molecules. Also, it should be pointed out that different types of ionizing particles have different efficiencies for producing excited solvent molecules. And further, it should be remembered that only a small fraction of the total particle energy goes into the production of excited solvent molecules; only about 4-6% of electron kinetic energy, about 0.5-0.7% of alpha particle kinetic energy, and about 1.0% of proton kinetic energy.

Primary Excitation Process

The ionizing particles interact with molecules (mainly solvent molecules) as they are slowed down and finally stopped in the scintillator solution. The kinetic energy is released in many forms. The bulk is converted into thermal energy (kinetic) of the molecules. Other interaction products include:

S^*	excited molecules
$S^+ + e^-$	ions and electrons
$A^{\cdot}$	free radicals
B^+, C^-	ion fragments
D, E, F	molecular fragments

4

The concentration of these species can be great along the primary track of the particle. The specific ionization of the particle will determine the concentration. The high specific ionization of alpha particles leads to high concentrations of the various products. Because of the high concentration, many of the excited molecules and ions interact with the other products leading to a reduction of excited solvent molecules and the subsequently lower photon yield. This type of quenching is often referred to as "track" quenching. Track quenching is less for the lower specific ionization electrons (i.e., beta particles). In some cases, secondary electrons produce excited solvent molecules.[1]

The numbers of solvent ions and excited solvent molecules are both important in the determination of the scintillation yield. Ion recombination can lend to an appreciable fraction of the number of excited solvent molecules which lead to the production of photons. Previous studies (2-4) showed that in some solvent systems, 60% of the observed fluorescence was the result of ion recombination. Essentially, all of the excited solvent molecules produced by the ionizing particle are excited to upper excited energy levels ($n \geq 2$). These upper excited energy levels undergo an internal conversion process (non-radiative) to produce the first excited singlet state. It is the yield of the first excited singlet state of the solvent molecules which determined the maximum scintillation yield.[5] Some studies have shown that energy transfer involving upper excited energy levels can occur when the energy acceptors (solutes) are present in high concentrations.[3] One reason for the different efficiencies of solvents is the fact that the internal conversion efficiency from upper excited states to the first excited state are different. Table 1 summarizes the known data on the calculated and measured relative efficiencies of some aromatic solvents.

Energy Migration

The excitation energy migrates from one solvent molecule to its neighbor solvent molecule. In this manner, the energy moves from one area to another until the solvent gives its excitation energy to other molecules in the liquid scintillator system. (These other molecules can be scintillator solutes or quencher molecules.) Two theories have been presented to describe the energy migration processes.

TABLE I. Comparison of calculated excitation yield from ion recombination and direct excitation with experimentally measured values for common liquid scintillation solvents.

Solvent	Internal Conversion (I.C.) Efficiency (a)	Direct Excited Solvent Yield	Excited S* Yield From I.C.	Excited S* Yield From Ion Recombination	Total Yield	Normalized Yield (%)	Measured(b) S_x (%)
Benzene	0.44	0.9	0.4	1.2	1.6	84	85
Toluene	0.76	0.9	0.7	1.2	1.9	100	100
p-Xylene	1.00	0.9	0.9	1.2	2.1	111	110
1,2,4-Tri-methyl Benzene	1.00	0.9	0.9	1.2	2.1	111	112

(a) C. W. Lawson, F. Hirayama and S. Lipsky, J. Chem. Phys. 51, 1590 (1969).

(b) At 5 g/l of PPO

Birles[16] described the transfer as being due to the formation and disassociation of two solvent molecules to form excited dimers (excimers). In this process, the energy may be transferred to the previously unexcited solvent molecule when the excimer breaks apart. Voltz[7] stated that the energy actually jumps from the excited solvent molecule to its neighbor by a non-radiative process. Energy migration can lead to transfer between many solvent molecules before actual transfer to solute molecules.

Energy Transfer

Because most scintillator solvents have properties which reduce the yield of photons, often molecules are added which efficiently accept the solvent excitation energy and emit that energy as photons. The efficiency of scintillator solutions is dependent upon how efficiently the energy is scavanged by these added molecules (solutes). Some of the properties of solvents which make them poor scintillators by themselves are:

a. Solvent molecules have low probabilities for photon emission.

b. The energy (wavelength) distribution of emitted photons is in the range where common detectors (multiplier phototubes) have reduced sensitivity.

c. The emission lifetimes are long ($\sim$30 nanoseconds) which means a greater probability of quenching before emission.

d. Due to the high solvent concentration, the probability of reabsorption of emitted photons is high.

The properties of the solute molecules are such as to minimize these drawbacks. The solute molecules have:

a. High fluorescence probabilities, $\sim$90%.

b. Wavelength distributions which match favorably with peak sensitivity of MPTs.

c. Very prompt photon emission, lifetimes between 1-2 nanoseconds.

7

d. Very low reabsorption probability because solutes
    are present in low concentration.

The transfer of energy from excited solvent molecules to
acceptor molecule (solute or quencher) is considered to be
basically a long range interaction and is not diffusion-
controlled.[8] At fairly low solute concentrations ($\sim 10^{-2}$ M),
the energy transfer process is quantitative[9] with many
solutes. This means that every excited solvent molecule
leads to an excited solute molecule. At lower concentrations,
the energy transfer efficiency decreases with a corresponding
decrease in photon yield. The energy transfer from solvent
to solute is not reversible because of a vibrational de-
excitation in the solute molecule, leaving it with insufficient energy to re-excite a solvent molecule. Thus, the
excitation energy is trapped by the solute molecules.

A second solute is sometimes used in liquid scintillation
counting. In early times, the second solute was used to
shift the spectral distribution of photons to more closely
match the most sensative response range of the MPTs.   In
more recent times, with the new bi-alkali MPTs, the secondary
solute is used more to reduce the effect of certain "color"
quenchers which may be present in the scintillator-sample
system. The concentration of the second solute can be
adjusted to provide quantitative energy transfer from the
first solute to the second solute.[10] Again, the molecular
internal de-excitation of the second solute renders the
energy transfer irreversible. Usually, the concentration of
the second solute is only a few percent of the concentration
of the first solute.

The energy transfer processes are all non-radiative, i.e.,
no photons are emitted and then reabsorbed by other molecules.
This has been most dramatically demonstrated by measurements
of the fluorescence decay times of liquid scintillator
solutions. Excitation of the solvent molecules alone showed
no change in the decay time of solute fluorescence compared
to direct excitation of the solute molecules. Thus, the
energy transfer processes must be many times faster than the
decay times of the solutes.[10]   Energy transfer processes
are of the order of 10^2 to 10^3 times faster than the fluorescence decay time of the fastest known solute ($\sim 10^{-9}$ seconds).

A recent experiment[10] using the pulsed Van de Graaff
showed the energy transfer from primary solute (PPO) to
secondary solute (M_2-POPOP) had an energy transfer rate

constant of about 2.5×10^{13} 1/mol-sec., which is about 10^3 times greater than the diffusion-controlled rate constant. It was possible to measure the relative emission of PPO and M_2-POPOP with constant PPO concentration and increasing M_2-POPOP concentration because of the high intensity of scintillation produced by the Van de Graaff pulses. Figure 2 shows the fluorescence spectra obtained.

Quenching Mechanisms

Quenching is a term commonly used to denote some process which causes a decrease in the photon yield of a liquid scintillator solution relative to no quenching (i.e., ideal). These quenching mechanisms can be divided into four main categories:

    a.   Energy absorption.

    b.   Dilution - concentration.

    c.   Impurity (sometimes called chemical).

    d.   Color.

The first category, energy absorption, has been discussed in the section on "Interaction of Ionizing Radiation", and includes wall effect, self-absorption, dilution-absorption, and annihilation. All of these processes involve the release of part or all of the particle energy to non-scintillation-producing media.

Dilution-concentration refers to the reduced efficiency of energy migration and energy transfer processes. A second solvent can dilute the scintillation solvent to a degree that interferes with the normal energy migration process. Thus, the excitation energy does not have the ability to migrate to the regions of the solute molecules. If the concentration of the solute(s) is reduced below the ideal concentration, there will be insufficient solute molecules to efficiently scavange the excitation energy from the excited solvent molecules.

Impurity quenching refers to the competition of other non-fluorescent molecules for the solvent excitation energy. These non-fluorescent molecules will prevent the energy from being transferred to the solute molecules.

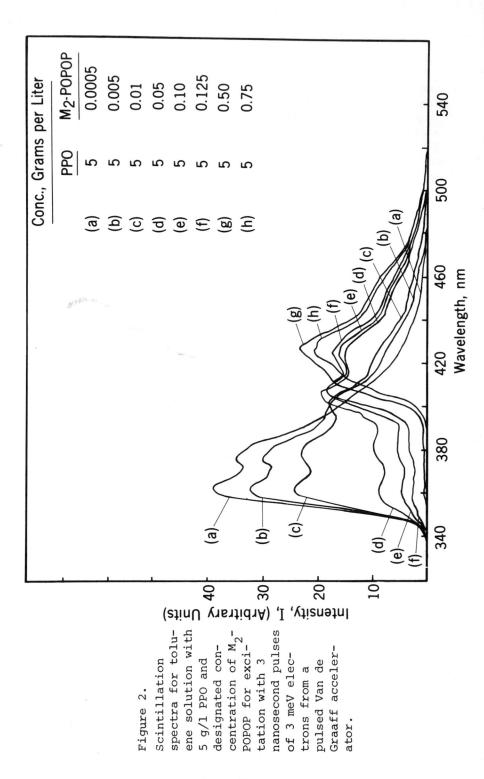

Figure 2.
Scintillation spectra for toluene solution with 5 g/l PPO and designated concentration of M_2-POPOP for excitation with 3 nanosecond pulses of 3 meV electrons from a pulsed Van de Graaff accelerator.

	Conc., Grams per Liter	
	PPO	M_2-POPOP
(a)	5	0.0005
(b)	5	0.005
(c)	5	0.01
(d)	5	0.05
(e)	5	0.10
(f)	5	0.125
(g)	5	0.50
(h)	5	0.75

These three processes all decrease the number of excited solute molecules formed and thus the number of photons released. The latter process, color quenching, refers to molecules present in the scintillator solution which absorb the photons after they have been produced.

There are three other factors which can lend to a reduced response by reducing detection of photons from the scintillator solution:

   a.   Photon trapping in scintillation vial.

   b.   Photon loss in optical light collection system.

   c.   Quantum efficiency of the MPTs.

Glass counting vials prevent a small fraction of the produced photons from escaping from the scintillation solution by reflecting the photons at the outside of the glass wall, at the air-glass interface, due to the difference in the index of refraction of glass and air. Since the index of refraction of toluene and glass are nearly the same, no light is lost[11] at the inside of the glass vial wall. Gordon and Curtis showed that the introduction of rough areas on the outside wall of a glass vial increased the photon escape from a liquid scintillation vial by disrupting the internal reflection process. This effect is more pronounced for the measurement of events which produce photon yields near the threshold of detection. Roughing the outside of a glass counting vial can increase the tritium counting efficiency up to 2%.

The optical light collector is that part of the counting system which guides the emitted photons onto the face of the MPTs. Highly efficient diffuse reflector material gives the most efficient means of collecting and guiding the photons. This property is usually fixed by the instrument, but can sometimes change if for any reason the diffuse reflector has been contaminated or has gotten dirty. This is important for systems which are very old or have had spillage in the counting chamber.

The quantum efficiency of the MPTs is also an instrument-fixed factor. Most commercial instruments use MPTs with bi-alkali photo cathodes which have quantum efficiencies of

TABLE 2. Summation of scintillation effiency parameters as function of electron energy

Electron Energy keV	Sx	N̄ph	Photons per keV	keV/photon	keV/photoelectron
1000	0.062	19,375	19.4	.052	.186
500	.062	9,688	19.4	.052	.186
300	.062	5,813	19.4	.052	.186
158	.058	2,864	18.1	.055	.196
50	.052	813	16.3	.061	.218
18.6	.047	273	14.7	.068	.243
5	.040	63	12.6	.079	.283
1	.031	9.7	9.7	.103	.368
0.5	.024	3.8	7.5	0.133	.475

about 28% for photons of 380–400 nm wavelength. The MPTs are usually operated under conditions such that a single electron produced at the photocathode which undergoes the full amplification will, on the average, produce a measurable pulse.

Scintillation Efficiency

Scintillation efficiency is defined as the energy released [12] as photons divided by the energy of the ionizing particle. The scintillation efficiency varies with energy of the particle. Figure 3 shows a plot of scintillation efficiency (Sx) as a function of the energy of electron. Above 300 keV energy, the scintillation efficiency of this liquid scintillator solution (5 g PPO and 0.1 g M_2-POPOP per liter of toluene) was independent of electron energy. Below 300 keV, the scintillation efficiency decreased with decreasing electron energy. Thus, 1 keV electrons produced only one-half as many photons per unit energy as 300 keV electrons. Table 2 summarizes the values of scintillation efficiency (Sx), average number of photons ($\bar{N}_{ph}$), photons per keV, and keV per photon of a scintillator solution (5 g PPO and 0.1 g M_2-POPOP in one liter of toluene) excited by different energy electrons. The last column is based upon a MPT which has an average quantum efficiency of 28% for photons from the liquid scintillator solution.

Table 3 and Figure 4 give the same information for excitation of the same liquid scintillator solution with alpha particles of different energy. The scintillation efficiencies for alpha particles is about one-tenth the scintillation efficiency for electrons.

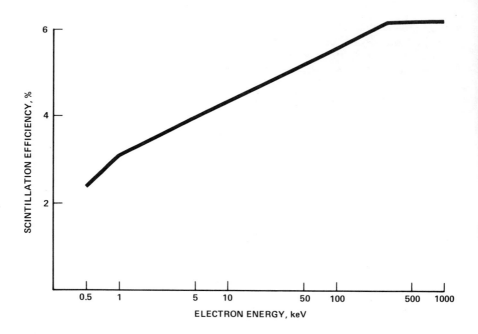

Figure 3.   Scintillation efficiency as a function of
electron excitation energy.

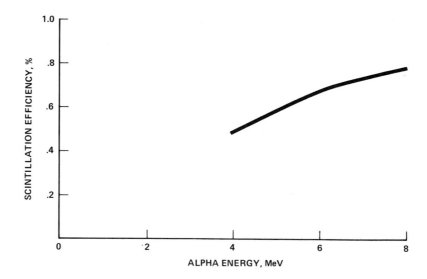

Figure 4   Scintillation efficiency as a function of
alpha particle excitation energy.

TABLE 3. Summation of scintillation efficiency parameters as function of alpha particle energy

Alpha Particle Energy MeV	Sx	$\bar{N}ph$	Photons per keV	keV per Photon	keV per Photoelectron
8.0	0.0078	19569	2.45	0.41	1.46
7.5	.0076	17922	2.39	.42	1.50
7.0	.0074	16178	2.31	.43	1.54
6.5	.0071	14338	2.21	.45	1.61
6.0	.0068	12788	2.13	.47	1.68
5.5	.0064	10947	1.99	.50	1.79
5.0	.0058	9106	1.82	.55	1.96
4.5	.0054	7556	1.68	.59	2.11
4.0	.0049	6103	1.53	.65	2.32

15

References

1. G. Laustriat, in "Organic Scintillators", D. L. Horrocks, ed. (Gordon and Breach, New York, 1968) pp. 127-145.

2. R. Cooper and J. K. Thomas, J. Chem. Phys. $\underline{48}$, 5097 (1968).

3. D. L. Horrocks, J. Chem. Phys. $\underline{52}$, 1566 (1970).

4. J. B. Birks, in "Organic Scintillators and Liquid Scintillation Counting", D. L. Horrocks and C. T. Peng, eds. (Academic Press, New York, 1971) pp. 3-23.

5. C. W. Lawson, F. Hirayama and S. Lipsky, J. Chem. Phys. $\underline{51}$, 1590 (1969).

6. J. B. Birks, J. C. Conte and G. Walker, IEEE Trans. Nucl. Sci. $\underline{NS-13}$, No. 3, 148 (1966).

7. R. Voltz, G. Laustriat and A. Coche, C. R. Acad. Sci., Paris $\underline{257}$, 1473 (1963).

8. Th. Forster, Discuss. Faraday Soc. $\underline{27}$, 7 (1959).

9. I. B. Berlman, J. Chem. Phys. $\underline{33}$, 1124 (1960).

10. D. L. Horrocks, Nucl. Instr. Meth. $\underline{128}$, 573 (1975).

11. B. E. Gordon and R. M. Curtis, Anal. Chem. $\underline{40}$, 1486 (1968).

12. D. L. Horrocks, "Applications of Liquid Scintillation Counting" (Academic Press, New York, 1974) pp. 27-30.

LIQUID SCINTILLATION ALPHA COUNTING AND SPECTROMETRY AND ITS APPLICATION TO BONE AND TISSUE SAMPLES*

W. J. McDowell
Chemical Technology Division
Oak Ridge National Laboratory
Oak Ridge, Tennessee 37830

and

J. F. Weiss
University of Tennessee
Comparative Animal Research Laboratory
Oak Ridge, Tennessee 37830

ABSTRACT

Three methods for determination of alpha-emitting nuclides using liquid scintillation counting are compared, and the pertinent literature is reviewed. Data showing the application of each method to the measurement of plutonium concentration in tissue and bone samples are presented. Counting with a commercial beta-liquid scintillation counter and an aqueous-phase-accepting scintillator is shown to be accurate only in cases where the alpha activity is high (several hundred counts/min or more), only gross alpha counting is desired, and beta-gamma emitters are known to be absent from the sample or present at low levels compared with the alpha activity. Counting with the same equipment and an aqueous immiscible scintillator containing an extractant for the nuclide of interest (extractive scintillator) is shown to allow better control of alpha peak shift due to quenching, a significant reduction of beta-gamma interference, and, usually, a lower background. The desirability of using a multichannel pulse-height analyzer in the above two counting methods is stressed. The use of equipment and procedures designed for alpha liquid scintillation counting is shown to allow alpha spectrometry with an energy resolution capability of 200 to 300 keV full-peak-width-at-half-peak-height and a background of 0.3 to 1.0 counts/min, or as low as 0.01 counts/min if pulse-shape discrimination methods are used. Methods

---

*Research sponsored by the Energy Research and Development Administration under contract with the Union Carbide Corporation and the University of Tennessee Comparative Animal Research Laboratory. Work was performed at Oak Ridge National Laboratory.

17

for preparing animal bone and tissue samples for assay are described.

INTRODUCTION

Although the principal use of liquid scintillation counting is the assay of low-energy beta emitting nuclides, it has been known for more than two decades that alpha particles could be counted effectively by this technique (1). The use of liquid scintillation for alpha counting is attractive because of its 100% counting efficiency and simplicity of sample preparation; and, in the late 1950's and early 1960's, several workers have made applications of the method (2-6). Widespread use of alpha liquid scintillation counting has been hindered because of the high background count and very poor energy resolution for alpha energies associated with the usual beta liquid scintillation methods.

In 1964 and 1965 Horrocks (6) in this country and Ihle et al. (7) in Germany demonstrated that a useful degree of alpha energy resolution could be attained by liquid scintillation methods. However, little practical use was made of these high-resolution methods until recently because the detectors and associated electronics needed are not commercially available and because the nuclide must be placed in the scintillator as an organic-soluble complex. This latter requirement has, however, led to the use of liquid-liquid extraction as a convenient means for both isolating the desired nuclide and placing it in a water-immiscible scintillator solution. Several workers have developed liquid scintillation counting procedures for alpha-emitting nuclides based on liquid-liquid extraction separations and/or the use of a water-immiscible scintillator containing an extractant (extractive scintillator), both for use with beta liquid scintillation counters (2,6-10) and for the application of high-resolution alpha liquid scintillation spectrometry (11, 12). Even more recently, the introduction of pulse-shape discrimination electronics has allowed the separation of alpha pulses from beta- or gamma-produced pulses in liquid scintillation systems and, hence, has made dramatic reduction in background count in alpha counting possible (13, 14). Figure 1 shows a direct comparison between the usual beta liquid scintillation counter and the high-resolution detector as regards energy resolution and the separation of alpha pulses from beta-gamma pulses.

This paper will discuss and compare the various forms of liquid scintillation for alpha counting and their advantages and disadvantages. Because of the increasing

18

importance of studies of animal metabolism of alpha-emitting nuclides, particularly plutonium, and because of some studies of plutonium uptake currently in progress in our laboratories, liquid scintillation counting of plutonium in animal tissue and bone samples will be emphasized.

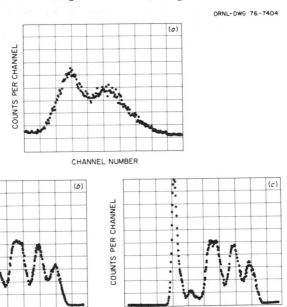

ORNL-DWG 76-7404

Fig. 1. Spectrum of ^{232}Th and daughters, ^{228}Th, ^{224}Ra, ^{220}Rn, and ^{216}Po with some ^{230}Th as seen (a) by a commercial β liquid scintillation counter and (b) by the high-resolution detector without β-γ rejection by pulse-shape discrimination and (c) with pulse-shape discrimination.

EXPERIMENTAL

Equipment

The commercial beta-liquid-scintillation detector used was a Packard Model 3214. An interface to a multichannel analyzer was arranged according to a diagram provided by D. C. Bogen (15) of the Health and Safety Laboratory, New York. Any of several commercial beta liquid scintillation counters could be used; and some types presently available have provisions for interfacing to a multichannel analyzer. It is desirable that the pulse-height information for the multichannel analyzer be taken from the scintillation counter circuitry at some point following the upper and lower discriminator and that it be linear with light output. This permits direct observation of the effect of adjusting the discriminators via the multichannel analyzer display and simplifies optimization of the settings to include all of the alpha peak while rejecting as much background as possible.

The high-resolution liquid scintillation alpha spectrometer was constructed from a detector consisting of an RCA 4523 or an EMI 9840A phototube sealed to a reflector-sample holder unit, as shown in Figure 2. Standard preamplifiers and linear amplifiers were used to present the pulse-height information provided by the phototube to a multichannel analyzer. Pulse-shape discrimination circuitry was assembled from standard components. More detailed descriptions of the high-resolution detector and associated electronics and of the pulse shape discrimination equipment are given in earlier publications (16, 17).

ORNL-DWG 75-3386

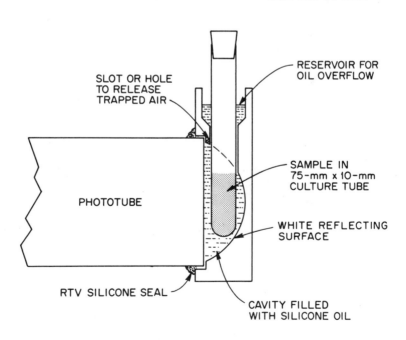

Fig. 2. Detector assembly for high-resolution alpha liquid scintillation spectrometer.

Reagents

The extractant used in the scintillator, di(2-ethyl-hexyl)phosphoric acid (HDEHP), was obtained from the Virginia Carolina Chemical Company. As received, this reagent contained about 1% impurities and was slightly colored. It was purified by a procedure involving precipitation of the copper salt of HDEHP. Details of this procedure are available in previous publications (18, 19). The tertiary amine, Adogen 364, was obtained from the Ashland Chemicals Company and was purified by single-stage high vacuum distillation. Naphthalene used in the scintillator was purified by resublimation or, for some tests, by zone refining. The scintillator, PBBO [2-(4'-biphenylyl-6-phenylbenzoxazole] was obtained from Eastman Chemical Products, Incorporated, and used without further purification.

Plutonium-239 was obtained from the Isotopes Division, ORNL. All other chemicals were of the usual reagent grade.

DESCRIPTION AND DISCUSSION OF PROCEDURES

Sample preparation

Small bone and tissue samples (< 25 g) are placed in solution with concentrated nitric acid and small amounts of 30% hydrogen peroxide. Gentle heat, to the entire container so as to prevent bumping, is provided primarily from radiant heaters. Digestion is continued until a clear solution is obtained. Samples with only small amounts of residual salts should not be allowed to go dry since plutonium and other nuclides may sorb on glass surfaces or otherwise become difficult or impossible to redissolve.

Larger-sized samples (> 25 g) are dissolved by successive treatments with concentrated nitric acid and 30% hydrogen peroxide, each treatment being followed by evaporation to dryness and heating to 450°C overnight in a furnace. If the residue is not white, the treatment with concentrated nitric acid and 30% hydrogen peroxide is repeated, followed by again heating overnight in the furnace. This series of steps usually produces a white ash, free of carbon and organic material. The ash is then dissolved in sufficient 2 M HNO_3, and the sample is subsequently treated by whatever method (outlined below) seems best according to the level of activity anticipated in the sample. Heating to 450°C does not cause a loss of plutonium in the larger-sized tissue samples, probably because of the presence of sufficient amounts of salts so that the plutonium enters the mineral

21

lattice rather than being adsorbed on the surface of the glass.

After the sample has been completely dissolved, the resulting clear nitric acid solution is treated in one of three ways, depending on the counting method to be employed: (a) If the sample is known to contain high levels of activity and is to be assayed using an aqueous-phase-accepting scintillator, the solution is diluted to a known volume after digestion and an aliquot is taken for addition to the scintillator. (b) If the sample is intended for direct extraction into a scintillator containing HDEHP, sufficient perchloric acid is added to the solution to give a final solution that will be 0.1 to 0.2 $\underline{M}$ in perchloric acid after all the metal ions present in the sample (from whatever source) have been converted to perchlorate salts. The nitric acid is evaporated with slightly increased heat (150 to 170°C), leaving only perchlorates and perchloric acid. This procedure ensures complete destruction of all remaining organic material and allows volume reduction to a salt solution containing a controlled amount of acid without the risk of the sample ever becoming dry. (c) If the sample is to be treated by an anion exchange separation (liquid or resin, see below), the solution is diluted or treated in such a manner as to yield the desired nitric or hydrochloric acid concentration.

Counting in a beta liquid scintillation counter using an all-purpose scintillator

Using a standard aqueous-phase-accepting scintillator, up to 1 ml of a highly salted aqueous phase such as that resulting from nitric acid dissolution of a tissue or bone sample can be incorporated directly into the scintillator. Such a sample can be counted in a commercial beta liquid scintillation counter. For gross alpha counting of samples with a sufficiently high count rate (e.g., at least several hundred counts/min), this counting method is adequate if measures are taken to ensure that the alpha peak is within the pulse-height range observed. Varying amounts of salt, acid, or even water cause the pulse-height response of the scintillator to change, thus shifting the position at which the alpha pulses appear. The peak position may be found by counting through a narrow window as the available pulse-height range is scanned, but the use of a multichannel analyzer connected to the counter is more convenient and also allows visual differentiation between alpha and beta-gamma spectra. Unless very similar amounts of salt, water, and acid are present in

each sample, the position of the alpha peak should be
monitored for each sample to ensure reproducible counting
efficiency.

Backgrounds of 20 to 30 counts/min plus a contribution
of 10 to 20 counts/min from ^{40}K in each gram of tissue are
usually encountered in alpha counting in this way. Some reduction of background count can be effected by careful
adjustment of the upper and lower discriminators to include
the alpha peak and reject as much of the beta-gamma contribution as possible; however, this advantage is gained at the
cost of a rather time-consuming individual discriminator
adjustment for each sample because of variable quenching.

The accuracy with which a sample can be counted by this
method is, of course, determined by the usual statistical
relationship that includes the background count; therefore,
the counting accuracy decreases rapidly as the net sample
count decreases toward, and falls below, the background count.
A total count equal to at least twice the background count is
usually considered the practical lower limit for counting
with reasonable accuracy. With samples prepared as above,
it must be remembered that ^{40}K and any other beta- or gamma-
emitting nuclides in the sample must be considered as background and in these cases background must be determined
individually for each sample from a spectrum of that sample
or it must be assumed that all the samples are alike in this
respect.

No energy resolution or pulse-shape discrimination is
possible with this counting method.

Counting in a beta-liquid-scintillation counter using an extractive scintillator

A scintillator containing 161 g of HDEHP, 80 g of naphthalene, and 4 g of PBBO (or 5 g of PPO) per liter of
toluene can be used to extract the plutonium (or most other
alpha-emitting nuclides) quantitatively from a properly
prepared sample. Recovery of plutonium from 5-ml samples
containing 1 g of dissolved bone was 100% for scintillators
containing 100 g or more of the extractant, HDEHP. Recovery
from tissue samples is easier than from bone and usually can
be done with lower HDEHP concentrations. This approach to
alpha-liquid scintillation counting, which has been used by
several workers (8-10, 12) has important advantages: (a) The
background contribution from ^{40}K in the sample is reduced
dramatically since HDEHP does not extract potassium and the
interaction of the ^{40}K beta with the scintillator is greatly
reduced. (b) The scintillator is more reproducible and the

pulse-height response to a given alpha is more nearly the same, thereby reducing the need for window adjustment for each sample. (c) Finally, and perhaps most importantly, an extracted nuclide can be stripped (see in following section) and reextracted into the high-resolution scintillator for analysis in the high-resolution detector, where energy resolution and much lower backgrounds can be attained.

The sample is prepared for extraction as outlined under "Sample preparation" [method b]. To facilitate plutonium extraction into HDEHP, 2 ml of saturated aluminum nitrate per gram of sample and sufficient water to make 5 ml per gram of sample is added to the sample while it is still hot.* A 5- to 8-ml volume of the sample is added directly to a 20-ml vial of the kind commonly used for beta liquid scintillation counting with 10 ml of extractive scintillator and the vial is shaken for 1 to 2 minutes. The phases are then allowed to separate, and the vial is placed in the counter. Counting can be done without separation of the aqueous phase. It has been shown by Keough and Powers (8), and Horrocks (11), that interphase counting efficiency varies from 2% for energetic betas to less than 0.1% for alpha, gamma, and low-energy beta emitters remaining unextracted in the aqueous phase. Although there is some quenching from extracted nitrate, there will be only minor variation in the pulse-height response of scintillator-samples prepared and extracted in this way. Although the pulse-height response is more con-sistent than with an aqueous-phase-accepting scintillator, it is still desirable to use a multichannel analyzer to monitor the position of the alpha peak.

Of twelve 400 to 800 count/min samples run as described above, the average plutonium recovery was 100.5% with a standard deviation of 1.2%.

Under the best conditions some alpha energy resolution is possible using this procedure. The full-peak-width-at-half-maximum-peak height (FWHM) is usually 0.9 to 1.0 MeV. Thus, it is sometimes possible to distinguish, and count separately, alpha energies that differ by more than 1 MeV. However, this procedure does not yield organic scintillator solutions sufficiently pure and free of quenching to allow their use in the high-resolution low-background detector.

Separation of alpha from beta-gamma pulses is usually not possible by this procedure either on an energy basis or

---

*Aluminum nitrate improves plutonium extraction by, (a) nitrate stabilization of Pu(IV), (b) complexing phosphate, and (c) increasing ionic strength.

via pulse-shape discrimination (20). Beta or gamma interactions with the scintillator produce light approximately
10 times more efficiently than do alpha particle interactions.
This difference in light-producing efficiency plus the fact
that both beta and gamma radiation produce a continuous
pulse height distribution from zero to some maximum value*
preclude separating this interference on an energy basis.

Although alpha and beta-gamma produced pulses are of
different pulse-shape (or pulse-time-duration) in some liquid
scintillator systems (13, 21), they cannot be separated on a
pulse-shape basis using presently available commercial beta-
liquid-scintillation equipment because the pulses are not
processed in the optical or electronic system in such a way
as to retain this information. Thus, although the total
background is less with the extractive scintillator used in a
commercial beta-liquid-scintillation counter (because ^{40}K and
some other beta-gamma emitters are not extracted), the background from outside sources cannot be reduced much. Background counts of 15 to 20 counts/min are the usual lower
limit.

Alpha spectroscopy with a high-resolution liquid scintillation detector

With appropriate detection and electronic equipment and
a properly prepared sample, it is possible to obtain an alpha
resolution of 200 to 300 keV FWHM. A background of about 1.0
count/min under an alpha peak can be obtained using a Pyrex
tube to contain the counting sample; this can be reduced to
0.3 counts/min using a quartz tube. Backgrounds of 0.01
counts/min or lower are easily obtained when pulse-shape discrimination methods are used to eliminate beta-gamma pulses.
Sample preparation methods for soil and water samples and
the necessary equipment for high-resolution alpha spectrometry have been described earlier (16), as has pulse shape
discrimination equipment for rejection of beta-gamma pulses
(13).

Sample preparation for high-resolution liquid scintillation: Optimum pulse height, energy resolution, and pulse-
shape discrimination require that the scintillator-sample
contain a minimum of color quenchers or chemical quenchers.
The extractant, the scintillator, and all other constituents

---

*In the case of beta radiation, because of the continuous
 beta energy distribution. In the case of gamma radiation,
 because of varying efficiency of gamma interaction with the
 scintillator.

of a high-resolution scintillator must be as pure as is practicable, the aqueous sample from which the activity is to be extracted must contain little or no nitrate or chloride (both chemical quenchers), and oxygen must be removed from the scintillator-sample. Methods for achieving such conditions are not as difficult as one might assume. In general, a single preliminary separation from gross impurities followed by extraction into the scintillator is sufficient for the purification. One such procedure using solvent extraction methods to isolate uranium and plutonium from soil and water samples has been described in a previous publication (16). Any of the precipitation, ion exchange, or solvent extraction methods that are used to separate alpha-emitting nuclides for analysis by surface barrier (or other) counting methods (22-24) can easily be adapted to isolation of the same nuclides for high-resolution liquid scintillation spectrometry.

An adaptation of a chloride-system ion exchange procedure (24) for preliminary purification of some bone and tissue samples intended for plutonium analysis was tested, but results were inferior to the solvent extraction procedure. Of 11 spiked samples run by this method, the average recovery of plutonium was 87.5% with a standard deviation of 7.6%.

The procedure that we have found to be most convenient and most reliable for sample purification for plutonium analysis in bone and tissue samples is an adaptation of the solvent extraction method described earlier (16). After the sample has been placed in solution in nitric acid and the acidity adjusted to 1 to 2 $\underline{M}$ by evaporation and dilution, aluminum nitrate is added to saturation and the sample is extracted with an equal volume of 0.3 $\underline{M}$ high-molecular-weight tertiary amine (octyl, nonyl or decyl) nitrate in toluene.* Just prior to extraction, the plutonium valence is adjusted to (IV) by the addition of sodium nitrite. After extraction, the aqueous phase is discarded and the organic phase is washed once with 0.7 $\underline{M}$ nitric acid. The plutonium is then stripped from the organic phase with an equal volume of 0.3 $\underline{M}$ $HClO_4$ followed by two washes of 1/4 to 1/5 the organic-phase volume containing a total of 0.3 to 0.5 meq. of $HClO_4$ and about 30 meq. of $LiClO_4$. In the stripping step the 0.3 $\underline{M}$ amine nitrate is converted to amine perchlorate, releasing an equivalent amount of nitric acid to the aqueous phase; thus, the stripping solutions must contain sufficient perchloric acid for this exchange plus that intended to remain in the sample. The strip solution is then heated to a

---

*Amines containing more than 0.1 to 0.5% primary and secondary amines are not suitable (16).

controlled temperature of 150 to 170°C in order to evaporate
the nitric acid and reduce the volume while retaining
perchloric acid (b.p. 200°C).* After the volume has been
reduced to approximately 5 ml, the solution is diluted,
while still warm, to 8 to 10 ml with a 2.5 w/v% solution of
sodium peroxysulfate. This final solution should be approximately 3 M perchlorate and 0.05 to 0.1 M in perchloric acid
with no other anions present.

For extraction into the scintillator, the sample is
quantitatively transferred to a conical separatory funnel
or centrifuge tube containing 1.2 to 1.5 ml of the high-
resolution extractive scintillator composed of 4 g/liter of
PBBO, 200 g/liter of naphthalene, and 64 g/liter of HDEHP in
toluene. After equilibration by manual shaking for 1 min
and phase separation, a measured quantity of the scintillator is transferred to a 10 x 75 mm culture tube, placed in
the high-resolution detector, and counted. The total
plutonium count in the original sample is then calculated
from the volume or weight ratio, scintillator taken/scintilla-
tor counted. Of 12 spiked samples reduced in volume, extracted, and counted in this way, the average plutonium recovery
was 101% with a standard deviation of 2.4%.

The excellent plutonium recovery by these methods
extends to low-level samples containing large amounts of
unwanted ions. In six samples of approximately 130 g in size
(half a rat), each containing a known 114 disintegration/min
of plutonium, the average recovery was 112 counts/min or
98.4% with a standard deviation of 4.1%. No interference
from naturally occurring radioactive elements was encountered.
These samples were prepared by the acid dissolution, furnace
heating and amine solvent extraction procedure outlined for
larger-sized samples under "sample preparation".

In all these procedures it is desirable that the volume
of the aqueous phase in the final extraction be about 5 to
10 ml since the amount of plutonium extracted depends not
only on the distribution coefficient of plutonium (concentration in organic phase/concentration in aqueous phase) but
also on the volumes of each phase. It should also be emphasized that careful analytical technique with attention to
quantitative transfers in all steps is required for good
results.

_____

*In the previous procedure (Ref. 16), fuming was continued
  until a lithium perchlorate fusion resulted. In some cases
  sufficient acid was removed to cause hydrolysis of the
  plutonium with attendant low recovery. The procedure
  described here is much superior.

When pulse-shape discrimination is to be used to reject the beta-gamma background, it is necessary to remove oxygen from the sample so that sufficient pulse-shape discrimination can be obtained. Previous publications have indicated that this would require freezing, evacuation, and helium or argon refilling of the sample tube (6, 16). Recent work has shown that excellent pulse-shape discrimination can be attained much more simply (17). Sparging with any of a number of inert gases, such as argon, carbon dioxide, or methane, for approximately 5 min produces a sample having excellent pulse-shape separation characteristics. Figure 3 shows beta-gamma and alpha pulse separation with such a sample and an illustration of the rejection of a beta-gamma background from an alpha spectrum. Samples prepared by this sparging method and closed with a cork covered with RTV silicon sealer usually remain stable and retain their good pulse-shape discrimination characteristics for more than 20 hr, some for many months.

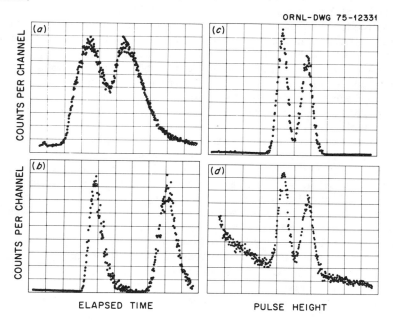

Fig. 3. Pulse-shape discrimination: Time spectra (a) before, (b) after argon bubbling; abscissa is 200 nsec full scale; β-γ pulses in left peak, α pulses right peak. Energy spectra; (c) pulse-shape discriminator rejecting β-γ pulses, (d) without pulse-shape discriminator (both peaks are α).

Stripping from extractive scintillator for reanalysis
in the high-resolution system: In some cases it has been
found desirable to first analyze all samples by counting in
the beta-liquid-scintillation counter using the extractive
scintillator and then reanalyze those requiring better energy
resolution or lower background with the high-resolution
system. In order to do this, the plutonium must be stripped
from the first extractive scintillator and reextracted into
the high-resolution scintillator. The following procedure is
used: The extracted sample is transferred from the 20-ml
vial to a 60-ml separatory funnel using small volumes of
toluene to wash. The aqueous phase is removed and discarded
and the organic phase is washed twice with 5-ml volumes of a
solution that is 0.4 $\underline{M}$ in LiCl and 0.1 $\underline{M}$ in HCl (to remove
nitrate from the organic phase). The washes are discarded.
A 10-ml volume of 0.2 $\underline{M}$ solution of 2,5-di-tert-butylhydro-
quinone in 2-ethylhexanol is added to the separatory funnel
[to reduce Pu(IV) and Pu(VI) to Pu(III)], and the plutonium
is stripped from the organic phase with two 10 ml volumes
of 6 $\underline{M}$ HCl. The combined HCl strip is then evaporated and
prepared for extraction as described in the previous section.
Eleven samples, each containing 1 gram of bone, treated in
this way gave an average recovery of 97% of the plutonium
added with a standard deviation of 1.7%.

CONCLUSIONS

Methods of alpha counting by liquid scintillation
methods have been reviewed and compared, and their successful
application to counting plutonium in bone and tissue samples
has been reported. A method of gross alpha counting using
scintillators and equipment designed for beta scintillation
counting in aqueous samples is shown to be useful where the
alpha count rate is sufficiently high and where contribution
to the background from beta or gamma emitters in the sample
is sufficiently low. A second method in which beta-liquid-
scintillation equipment is used but the plutonium is
extracted into a water-immiscible scintillator was found
to give more reproducible results and a lower background.
For samples with low count rates or those requiring alpha
energy resolution, an alpha-liquid-scintillation method
is found to offer advantages of very low background (about
0.01 counts/min), the ability to selectively count alpha
pulses while rejecting counts from beta or gamma emitters
in the sample, and an energy resolution capability of 200 to
300 keV FWHM. Sample purification and separation methods
based on solvent extraction (or on a combination of ion

exchange and solvent extraction) for these alpha liquid scintillation methods are more simple and more rapid than those for most other alpha counting methods.

Acknowledgements - The authors wish to acknowledge the excellent technical assistance of G. N. Case and Nancy Lovro.

REFERENCES

1. J. K. Basson and J. Steyn, Proc. Phys. Soc. (London) A67 297 (1954).
2. D. L. Horrocks and M. H. Studier, Anal. Chem. 30, 1748 (1958).
3. H. H. Seliger, Intern. J. Appl. Radiation Isotopes 8, 205 (1961).
4. L. E. Glendenin, Ann. N. Y. Acad. Sci. 91, 166 (1961).
5. K. F. Flynn, L. E. Glendenin, E. P. Steinberg and P. M. Wright, Nucl. Instr. Methods 27, 13 (1964).
6. D. L. Horrocks, Rev. Sci. Instr. 35, 334 (1964).
7. H. R. Ihle, M. Korayannis and A. P. Murrenhoff, in Proceedings of the Symposium on Radioisotope Sample Measurement Techniques, p. 485 (Vienna: IAEA) (1965).
8. R. F. Keough and G. J. Powers, Anal. Chem. 42, 419 (1970).
9. J. P. Ghysels, Ind. At. Spatiales 1, 36 (1972).
10. T. K. Kim and M. B. MacInnis, in Organic Scintillators and Liquid Scintillation Counting (edited by D. L. Horrocks and C. T. Peng), p. 925, New York: Academic Press (1971).
11. D. L. Horrocks, Intern. J. Appl. Radiation Isotopes 13, 441 (1966).
12. W. J. McDowell, in Organic Scintillators and Liquid Scintillation Counting (Edited by D. L. Horrocks and C. T. Peng), p. 937, New York: Academic Press (1971).
13. J. H. Thorngate, W. J. McDowell and D. J. Christian, Health Phys. 27, 123 (1974).
14. J. W. McKlveen and W. R. Johnson, Health Phys. 28, 5 (1975).
15. D. C. Bogen, private communication (1970).
16. W. J. McDowell, D. T. Farrar and M. R. Billings, Talanta 21, 1231 (1974).
17. J. W. McKlveen and W. J. McDowell, Nuclear Technology 22, 159 (1976).
18. J. A. Partridge and R. C. Jensen, J. Inorg. Nucl. Chem. 31, 2587 (1969).

19. W. J. McDowell, P. T. Perdue and G. N. Case, <u>J. Inorg. Nucl. Chem.</u> (in press).
20. J. W. McKlveen, private communication (1974).
21. D. L. Horrocks, <u>Applications of Liquid Scintillation</u> Counting, pp. 276-289 (New York: Academic Press) (1974).
22. C. W. Sill, K. W. Pauphal and F. D. Kindman, <u>Anal. Chem.</u> <u>46</u>, 1725 (1974).
23. N. Y. Chu, <u>Anal. Chem.</u> <u>43</u>, 449 (1971).
24. G. H. Coleman, <u>The radiochemistry of plutonium</u>, Report NAS-NS 3035 (1965).
25. P. T. Perdue, D. J. Christian, J. H. Thorngate and W. J. McDowell, <u>J. Inorg. Nucl. Chem.</u> (in press).

Measurement by Liquid Scintillator of Labelled Compounds (^{3}H or ^{14}C) Dropped onto Supports

Sonia Apelgot and Maurice Duquesne
Laboratoire Curie de la Fondation Curie - Institut du Radium
11 rue P. et M. Curie. 75231 Paris Cedex 05 (FRANCE)

Abstract

Radioactive compounds dropped onto supports are measured correctly, even if not extracted by liquid scintillator. Their dissolution is not required, it suffices simply to have a contact between the β particles and the liquid scintillator.

The difficulties encountered with paper support exist only in the case of ^{3}H (and not ^{14}C) and are a consequence of the paper structure itself. These difficulties disappear when the papers are counted wet (and not dried) in a liquid scintillator containing dioxane. Under these conditions, the measuring efficiencies are not very different from those obtained with glass fibres, or with a homogeneous phase; they no longer depend on the size of tritiated compound molecules.

Introduction

An unfavourable prejudice exists against measurement, by liquid scintillator, of labelled compounds dropped onto supports. The principal objection is that, under such conditions, the radioactive compounds dissolved in the liquid scintillator are counted with a greater efficiency than that of the insoluble compounds remaining on this support (1 and 2). But experiments have shown that these assertions are false.

The first supports used were of paper. Under these conditions, experiment shows that measurement efficiciencies are, in fact, low when the radioelement is ^{3}H, and the radioactive compounds are insoluble in the liquid scintillator. From these results, the unfavourable conclusions regarding the general use of supports were perhaps drawn - and these - probably, through analogy to the example of solutions of labelled compounds which are non miscible in the liquid scintillator. However, the studies performed in our laboratory demonstrated that it is the paper structure, and it alone, which is responsible for the results obtained with these supports; the soluble or insoluble characteristic of tritiated compounds does not play any role (3 and 4). By using glass fibre as a support, the measurement efficiencies are high, even with tritiated compounds insoluble in liquid scintillators.

Materials and Methods
  1. Apparatus: Automatic Spectrometer (Intertechnique, France).
  2. Liquid scintillators:
      a) water - non-miscible : NE211 (Nuclear Enterprises, G.B.) or solution prepared at the laboratory, containing toluene (Baker Chemical D.V.), PPO 4 g/l and dimethyl POPOP 0.1 g/l.
      b) water - miscible (with 10% water) : NE220 (Nuclear Enterprises, G.B.) containing dioxane.
We generally used either 1 or 3 ml of one or the other of these liquid scintillators.
  3. Supports:
      a) paper made with glass fibre (Whatman ref. GF/C or GF/A), called "glass fibre" in this work.
      b) Whatman paper n° 1 (185 mg/mm^2 or n° 17 440 mg/mm^2); these papers are made from cellulose.
  4. Radioactive solutions:
      a) standard solution of [^{3}H] thymidine (NEN, USA) or of [^{14}C] thymidine (CEA, France)
      b) bacteria of a thymineless E. Coli strain, labelled with ^{3}H; by using [^{3}H] thymidine, we obtained bacteria labelled exclusively in their DNA.
      c) [^{3}H] DNA: prepared from the previously labelled bacteria, according to the technique described by MARMUR (5).
  5. Measurement technique:
An aliquot of these radioactive solutions was dropped onto the selected support, which was measured:
      a) wet: as soon as prepared, the support was immersed in liquid scintillator NE220 containing dioxane. Under these conditions, the measured activity increased with time, to attain its equilibrium value in 3 to 6 hours.
      b) dry: the support was dried under an infra-red lamp, cooled and then immersed in a liquid scintillator without dioxane.
  6. Amplitude spectra:
The number of photons produced by a β particle of energy E_0, and consequently the number n of photoelectrons produced on the photocathode of a photomultiplier tube is, in proportion to the energy absorbed: $\bar{n} = KE$. This mean

value, $\bar{n}$, corresponds to a probability $p(n)$ of having n
photoelectrons by absorption of a β particle. The pulse
height spectrum at the anode of the photomultiplier tube
reflects this distribution. Thus it can be seen that for
each β disintegration there corresponds a value of $\bar{n}$ and a
mean amplitude $\bar{V}(n)$ of anode pulses. When $n = 1$, one
observed the spectrum S1E corresponding to a single photoelectron; when there is a large number of β disintegrations
the pulse height spectrum at the anode is the sum of the
spectra S1E, S2E,.... SnE; the mean value of the amplitudes
$\bar{V}_{exp}$, of the experimental spectrum permits deducing of the
mean value $\bar{n}$. The quenching phenomenon, and the poor transmission of photons caused by the paper's thickness, diminish
the value of the constant K in the previous formula $\bar{n} = KE$,
while the self-absorption phenomenon, in the paper, of the
β particles, diminishes the value of E_o. The net result of
these modifications is to diminish the values of $\bar{n}$ and is
manifested in a reduction of the anode pulse amplitudes.
Consequently, we observe an increase in the intensity of
spectrum S1E, to the detriment of that of the intensity of
the spectra S2E...SnE. The photomultipliers with porous
dynodes (RCA8850) permit the distinguishing of spectra S1E,
S2E and S3E, in an amplitude spectrum corresponding to the
detection of ^{3}H (Fig. la). In such spectra, quenching and
self-absorption manifest themselves by an increase in the
ratio $\dfrac{\text{Intensity S1E}}{\text{Intensity S2E}}$; that is, the ratio $P(1)/P(2)$. The
form of spectrum S1E, with a single photoelectron, allows
improving of the detection efficiency of a liquid scintillation apparatus (6).

Results
     An initial study was made, using solutions of tritiated
compounds and paper supports of different nature and thickness. These solutions, dropped onto supports, were dried,
then immersed in a liquid scintillator without dioxane. The
measurements of the radioactivity show that the efficiencies
(Table 1) are:
     - high, in the case of bacteria or DNA, though these
       compounds are insoluble in the liquid scintillator;
     - low in the case of thymidine; in this case, the paper
       thickness does not play any role.
     Therefore, such results show that the degree (high or
low) of measurement efficiency is independent of solubility
or insolubility of the tritiated compounds, in the liquid
scintillator. This efficiency depends neither on the nature
of the paper used for support, nor on its thickness.

TABLE I.

[³H] Samples dropped onto supports (paper or glass fibre) and counted dry

Samples	Activity (dpm)	liquid scintillator	n°	Whatman paper			Glass fibre			
				counted activity (cpm) with support (B)	support removed	E %	counted activity with support (A)	support removed	E %	B/A
[³H] T.D in H₂O	45·000	NE 211	1	2750, 2915, 2865 } 2840	12, 15, 15	6,5	20·520, 20·320, 20·555 } 20·465	445, 490, 690	45,5	0,14
		with Toluene	17	5695(*) → 2847, 2835	0, 0	6,5	38·880(*)→19·440, 20·290	185, 130	44	0,14
[³H] Bacteria in physiological serum	73·500	NE 211	1	25·445, 27·420 } 26·430	0	36	34·620, 32·275 } 33·445	120, 105	45,5	0,79
[³H] DNA in H₂O	25·890	with Toluene	1	7240, 7595 } 7415	0, 0	28,5	10·115, 9·975 } 10·045	65, 65	39	0.74

The [³H] thymidine ([³H] T.D) solution was a standard solution ; the activity of the [³H] bacteria suspension and the [³H] DNA solution was measured by the combustion technique. For each sample, we dropped 0.1 ml of labelled solution or suspension onto a support. They were dried and then immersed in 3 ml of liquid scintillator in the case of [³H] DNA, and in 1 ml in the case of the other radioactive solutions.
E represents the measurement efficiency.
(*) : for this sample, we immersed, in 3 ml of liquid scintillator, 2 papers, each containing 0.1 ml of the radioactive solution.

TABLE II

Counting of [^{14}C] thymidine on dry supports (glass fibre or paper)

Glass fibre		Whatman paper n° 1		
Counted activity (cpm)		Counted activity (cpm)		
with support (B)	support removed	with support (A)	support removed	B/A
40.065	1515	30.260	51	
40.180	1052	30.785	110	
39.795	3057	30.915	41	
average:		average		
40.015 ⟶ E=95%		30.655 ⟶ E=73%		0.77

The [^{14}C] thymidine solution was a standard solution; onto each support we dropped 0.05 ml, which corresponded to 42.120 dpm. Each support was dried and then immersed in 1 ml of a liquid scintillator without dioxane (NE211).

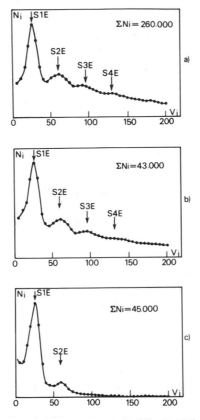

Figure 1: <u>Anode pulse amplitude spectra Ni (Vi)</u> (dry supports)

The same voluem of the tritiated standard solution was dropped either onto a paper (spectrum b) or onto a glass fibre (spectra a and c). Each support was dried then immersed in 1 ml of NE211. In the case of the C spectrum, we added 7.5% (v/v) of chloroform, to the sample thus giving the a spectrum. The amplitudes are given in arbitrary units, with the intensities normalised on the S1E spectrum. Σ Ni represents the total number of counted pulses.

Conversely, this degree of measurement efficiency depends on the size of tritiated compounds (thymidine or DNA). These phenomena do not exist, either with compounds labelled with ^{14}C (Table II) or with tritiated compounds dropped onto a glass fibre support (Table 1). The β self-absorption in the paper, or the paper's action on photon transmission, depend, on the contrary, on the thickness of the paper used and are independent of the size of the tritiated compounds' size and of the nature of the radioactive isotope. None of these phenomena is thus able to explain the results obtained: they are specific to ^{3}H and paper.

In order to understand the role of the paper, we studied the pulse height spectra of different samples prepared from a same radioactive solution, using, as a support, either paper or glass fibre; the radioactivity of each sample had been previously counted. The pulse height spectra were measured with the special apparatus described in paragraph n° 6 above (Materials and Methods).

For samples of ^{3}H on paper or glass fibre supports, experiment shows that the pulse height spectra are only slightly modified (Fig. 1a and 1b). The spectral shift of anode pulses to lower amplitudes were previously observed by FURLONG (7). This displacement is expressed by a lowering in the detection efficiency of β particles. To calculate this, we compared, on our spectra, the probability ratios $P(1)/P(2)$, $P(1)/P(3)$, and $P(2)/P(3)$. This calculation provides a detection efficiency of 81%, when the support is glass fibre (Fig. 1a), and 72% when it is paper (Fig. 1b). This decrease of 15% is not at all compatible with that of the counting rate which, for the same samples, differed by a factor of 1/7 (Table III).

In order to interpret these results, we continued the experiments by adding a quantity of chloroform to the sample containing the ^{3}H dropped onto glass fibre; the volume of chloroform added was such that the drop in the counting rate was similar to the case of paper supports. In the presence of chloroform, the spectral curve (Fig. 1c) is very different from that obtained in the presence of paper (Fig. 1b); the curve shows a definite shift towards the lower amplitude of the S1E spectrum which, this time, accounts perfectly for the observed drop in the counting rate (Table III). It is known that chloroform decreases the detection efficiency by lowering the value of the constant K (p. 3).

Thus, in the presence of paper or of a quantity of chloroform, both of which provoke a similar decrease in the counting rate, the detection efficiencies are different: with chloroform, the counting rate diminishes like the β detection efficiency while, with paper only the counting rate falls, the β detection efficiency itself being slightly modified (Fig. 1). It is thus a phenomenon of "all or nothing" which can be explained by assuming that a large number of [^{3}H] molecules are not "registered" by the liquid scintillator. This takes place as if 6 out of 7 β particles were "masked" by the paper, the 7th being correctly detected. For ^{3}H carried by large molecules, DNA, or ^{3}H incorporated in bacteria for example, about 75% of the β are detected (Table 1 and III).

It seemed to us that the difference between the results obtained using glass fibre and those obtained with paper, might be explained by the difference in structure of the supports.

With the aid of documentation generously provided to us by the "Ecole Française de Papeterie" at Grenoble, we were able to formulate the following hypothesis: the paper fibre behaves like a network of capillary tubes, permeable only to aqueous solutions; the glass fibres act like a network of impermeable threads, capable of retaining the diverse molecules only on their surface. In the capillary micro-structure of the paper, aqueous solutions carry with them only solutes of small size, radioactive or not, large-sized solutes remaining on the surface. When the radioelement is ^{3}H, the labelled compounds of small size are not detected for two reasons: firstly, because the liquid scintillator cannot penetrate the capillary microstructure and is never in contact with these molecules; and secondly, because the β particles from the ^{3}H have a path length which is too short to reach the liquid scintillator. These difficulties do not occur with large-sized molecules remaining on the surface, and thus in contact with the liquid scintillator; nor do they occur with the ^{14}C, whose β particles has more energy (Table II).

This permeability of the capillary microstructure of paper only to aqueous solutions should permit, as experiments have confirmed, better counting efficiencies when this support is measured wet. Under such conditions, a liquid scintillator containing dioxane is necessary, and it was found that ^{3}H is measured with an acceptable efficiency-about 80% - of that obtained with a glass fibre support (Table IV); moreover, this efficiency is similar to that obtained in the case of bacteria or DNA (Table 1). When these supports are dried and immersed in the liquid scintil-lator, it was observed that measurement efficiencies begin very low and increase with time. Under these conditions, the measured activity slowly approaches its equilibrium

TABLE III

Detection Efficiency of the β of ^3H

Supports	Activity measured (cpm)			R	
	M	I	I/Ir	%	R/R$_o$
– HCl$_3$ glass fibre	254.000	260.000	Ir	81	R$_o$
+ HCl$_3$	38.000	45.000	0.17	19	0.23
paper	36.000	43.000	0.16	72	0.89

0.1 ml of a [^3H] thymidine solution was dropped onto a paper or glass fibre disc. The discs were dried and then immersed in 1 ml of the liquid scintillator NE211. Each sample was first counted in the Intertechnique spectrometer apparatus, then in the apparatus having a photomultiplier with a porous dynode, which permits the registration of the anode pulse amplitude spectra. "M" represents the result of the measurement on the Intertechnique apparatus, and "I" the total number of the counted pulse serving to establish the amplitude spectra; "R" is the efficiency detection of the β calculated from the spectra, as explained in the text.

value (in 7 to 39 days), and in the case of the [^{3}H] thymidine used, the measured activity is seen in the liquid scintillator. This experiment demonstrates the slow diffusion of liquid scintillators, containing dioxane, in the interstices where water penetrates easily and where liquid scintillators without dioxane (such as NE211) have no access.

Discussion

The study of scintillation spectra demonstrated that paper does not significantly alter the scintillation phenomena (Fig. 1) and that it decreases the measurement efficiencies by only about 15% (Table III). Previous experiments have shown that the difficulties encountered with paper in measurements of ^{3}H activity were not related to its constituent – cellulose – but to its tubular structure, since these difficulties disappear with the use of electrophoretic membranes, whose only difference from paper lies in its cellulose structure (4). These previous results are confirmed by the present study, since we demonstrated that all these difficulties occur only when the measurements are carried out with dried paper. Paper has a multitude of microholes, and the results obtained can be explained if one assumes that these holes are permeated only by aqueous solutions. This paper specific structure explains that ^{3}H counting results are of the same kind, regardless of the nature, origin or even thickness of the paper (Table 1 and ref. 8). By measuring these paper supports wet, in a liquid scintillator containing dioxane, the measuring efficiencies have the same magnitude for all tritiated compounds, whatever their size, and they are very much the same as those obtained with glass fibres. Paper thus loses its "peculiarity" – meaning that the [^{3}H] β particles are detected independently of the tritiated compounds' size; the 20% decrease in the counting rate, which was observed, meshes with the 15% decrease in efficiency detection-deduced from analyses of the scintillation spectra. This paper reduction of the counting efficiency is the same as with ^{3}H or ^{14}C (Table 1, II and IV).

This study also demonstrated that the tritiated compounds dropped onto supports are, like the [^{14}C] compounds, correctly measured, even when not extracted by the liquid scintillator (Table 1 and II); their dissolution is not necessary and a contact only between the β of ^{3}H and the liquid scintillator is required. Each time it is possible to establish such a contact, the counting efficiencies are acceptable. Thus it is possible to measure ^{3}H or ^{14}C contained in bacteria or organs, without destroying them.

TABLE IV

Measurements in a liquid scintillator containing dioxane (NE 220)

Volume dropped (ml)	Measurements with wet supports					Measurements with dry supports						
	Glass fibres		Paper			Glass fibres		Paper				
	with support (a_1)	support removed	with support (b_1)	support removed	b_1/a_1	with support (a_2)	support removed	with support $t=0$ (b_2)	support equilibrium (b'_2)	support removed	b_2/a_2	b'_2/a_2
0.050	11·690	10·380	9·350	9·050	0.73	12·340	11·885	1.105	8.285 (+)	7.560	0.09	0.67
0.10	23·420	21·960	19·200	18·515	0.83	24·625	23·370	2.360	14·315 (+)	13·015	0.10	0.58
0.15	33·030	31·080	27·505	25·170	0.74	35·480	31·075	3.705	20·800 (●)	18·420	0.10	0.59
0.20	43·030	37·845	36·895	33·980	0.86	45·805	41·850	4.900	32·635(*)	30·015	0.11	0.71

We utilized an aqueous solution of [³H] thymidine. 0.050 or 0.10 ml were dropped onto support, the other volumes (0.15 and 0.20) onto two supports. The paper utilized here was Whatman n° 17. The measurement of radioactivity made at t = o corresponds to that made in the hour following the immersion of the support in NE 220; the equilibrium corresponds to t= 7 days for measurements marked (+); t = 19 days marked (●) and t = 30 days marked (*). Activities are given in cpm.

43

The cellular membranes are permeated by the solvents (toluene, xylene) common to all liquid scintillators. Since the compounds inside the cells are not very concentrated, they only slightly modify the scintillation processes. Counting of isolated cells is accomplished with a high degree of efficiency, even in the case of ^{3}H (4). Conversely the compact structure of a biological tissue hinders the scintillation phenomena. However, it suffices to destroy this structure through the use of an enzyme - pronase - to take correct measurements again (8 and 9). As the liquid scintillators pass through the cellular membranes, they extract the cells' soluble compounds and, due to this, become modified. This is the classical "quenching", which is determined by the usual methods. According to the nature of the labelled compounds inside the organs' cells, the samples to be counted, contain radioactive compounds dis-solved by the liquid scintillator, and others still "in situ", inside the cells. Experiments show that they are both counted with the same efficiency, that corresponding to the liquid scintillator being modified or not by the compounds, it will have dissolved (8 and 9).

This situation is different from that which exists in another kind of heterogeneous phase measurement, produced by 2 non-miscible liquids. In this case, the solutes are, in fact, shared between the 2 solvents, which form the two phases. The soluble solutes fraction in the liquid scintil-lator is counted with a greater efficiency than the insolu-ble fraction in the liquid which, therefore, remains dis-solved in its original solvent. This solvent, which is non-miscible in the liquid scintillator, forms a screen between the latter and the radioactive molecules. Thus, a defective contact exists between the β particles and the liquid scintillators.

Conclusions
    The compounds labelled with ^{3}H or ^{14}C and dropped onto supports, are correctly counted each time a contact is possible between the liquid scintillator and the β particles, even if the labelled compounds are insoluble in these scintillators. In the case of ^{14}C, such a contact exists, whatever the nature of the support is; in the case of ^{3}H, this contact is possible only if the support is of glass fibre, or if the support, when paper, is measured wet.

Acknowledgements
    We wish to thank Professor J. Chiaverina (Ecole Fran-çaise de Papeterie at Grenoble) for the documents and

information he kindly furnished to us, and the Societe Intertechnique, which measured for us some [^{3}H] DNA, and [^{3}H] bacteria samples, by combustion technique.

Bibliography
1. Y. Kobayashi and D.V. Maudsley in "Liquid Scintillation Counting: Recent Developments", p. 189-205 (P.E. Stanley and B.A. Scoggins, Eds.) Academic Press Inc., N.Y. (1974).
2. B.W. Fox, Symposium on "Liquid scintillation counting", Bath (Angleterre) (16 - 19 Sept. 1975) proceedings in press.
3. S. Apelgot and N. Rebeyrotte, Unpublished results.
4. S. Apelgot and M. Duquesne, J. Chim. Phys. 58, 774 (1961).
5. J. Marmur, J. Mol. Biol. 3, 208 (1961).
6. M. Duquesne, C.R. Acad. Sci. Paris, 267, 1347 (1968).
7. M.D. Furlong in "The Current Status of Liquid Scintil-lation counting", p. 201 (E.D. Bransome, Ed.) Grune and Stratton, New York and London (1970).
8. S. Apelgot, R. Chemama and M. Frilley, Bulletin du Cancer 60, 41 (1973).
9. S. Apelgot, R. Chemama and M. Frilley in "Liquid scintillation counting: Recent developments", p. 249-265, (P.E. Stanley and B.A. Scoggins, Eds.) Academic Press Inc., N.Y. and London (1974).

HETEROGENEOUS COUNTING ON FILTER SUPPORT MEDIA

E. Long, V. Kohler and M.J. Kelly
Beckman Instruments Inc.
Scientific Instruments Division
Irvine, CA 92713

Abstract
    Many investigators in the biomedical research area have
used filter paper as the support upon which radioactive
samples are counted. This means that a heterogeneous counting of sample sometimes results. The count rate of a sample
on a filter will be affected by positioning, degree of
dryness, sample application procedure, the type of filter,
and the type of cocktail used. Positioning of the filter
(up or down) in the counting vial can cause a variation of
35% or more when counting tritiated samples on filter paper.
    Samples of varying degrees of dryness when added to the
counting cocktail can cause nonreproducible counts if
handled improperly. Count rates starting at 2400 CPM initially, can become 10,000 CPM in 24 hours for ^{3}H-DNA (deoxyribonucleic acid) samples dried on standard cellulose acetate
membrane filters. Data on cellulose nitrate filters shows a
similar trend.
    Sample application procedures in which the sample is
applied to the filter in a small spot or on a large amount
of the surface area can cause nonreproducible or very low
counting rates. A tritiated DNA sample, when applied topically, gives a count rate of 4,000 CPM. When the sample is
spread over the whole filter, 13,400 CPM are obtained with a
much better coefficient of variation (5% versus 20%).
    Adding protein carrier (bovine serum albumin-BSA) to
the sample to trap more of the tritiated DNA on the filter
during the filtration process causes a serious beta absorption problem. Count rates which are one-fourth the count
rate applied to the filter are obtained on calibrated runs.
    Many of the problems encountered can be alleviated by a
proper choice of filter and the use of a liquid scintillation
cocktail which dissolves the filter. Filter-Solv has been
used to dissolve cellulose nitrate filters and filters which
are a combination of cellulose nitrate and cellulose acetate.
Count rates obtained for these dissolved samples are very
reproducible and highly efficient.

Quantitative results can only be obtained by dissolving
or emulsifying the filter and the sample in the cocktail.
The normal quench monitors of external standard channels
ratio or sample channels ratio are valid for dissolved
filters. When the filter is transparent or opaque in the
vial there are too many variables to be assured that a
stable, constant efficiency count rate is obtained.

Introduction

In pursuing biomedical literature these days, one is
struck by the relative frequency with which one finds that
the authors are counting samples on filter paper and measuring the disintegrations per minute (DPM) from these samples.
For instance, the book "DNA Synthesis In Vitro" (1) has
numerous articles on acid precipitated deoxyribonucleic acid
(DNA) counted on filter support media in which the counts
per minute are translated into specific activities. Recent
journal articles in such prestigious journals as Biochemistry
and Journal of Biological Chemistry have also contained
articles in which specific activities on filter support media
are claimed. All of these articles bring up the question as
to how are these counts as obtained in the liquid scintillation counter corrected for quench, self-absorption, and beta
absorption. Many authors appear to assume constant quench,
constant self-absorption, and constant beta absorption, as
well as other factors for each of their samples. This fact
prompted us to reiterate some previously known experimental
evidence on the position of the filter paper (2-4), the
solubility of samples in cocktails which leach the material
off of the filter paper and its inherent problems (5-9) and
the problem of reproducibility of sample spot size and the
use of carrier material to increase filtration efficiency.
These problems can be alleviated if the filter and solution
are able to be dissolved in a proper cocktail configuration.
In the subsequent discussion, we will present evidence to
show why it is necessary to dissolve the filter in order to
obtain the true activity of the sample and to give the magnitude of error encountered for each experimental condition
tried.

Filters and Cocktails

There are two general types of filters used in biochemical analysis. The first type is a "depth" or crude filter
which is used in routine separations. They are generally
made of coarse fibers which are pressed together to form
flow channels of random spacing and variable size. Filtered
samples are trapped throughout the matrix of the filters, and

due to the random orientation of the fibers, there is no absolute pore size. Paper filters and glass fiber filters are examples of this type.

The second type of filter is a "screen" or membrane filter that has definite ranges of particle retention. These filters are commonly composed of mixed esters of cellulose acetate, cellulose nitrate, and other polymers. These filters have a uniform microporous matrix. Millipore* and Sartorius** membrane filters are common representatives of this class.

A wide variety of cocktails have been used in counting radioactive material on filter papers. One of the most common cocktail formulations is a toluene/PPO cocktail formulation (6 grams PPO per liter of toluene) (7). Others have reported the use of similar toluene or xylene formulations with the presence of solubilizers or emulsifiers (Triton X-100****, Bio-Solv III***) (8,9).

There are two widespread uses for the depth and screen filters in radiochemical analysis. Many compounds of biochemical interest such as proteins and nucleic acids will precipitate out of solution in the presence of acids. Common acid solutions which cause this precipitation are: trichloroacetic acid (TCA) (10%), or perchloric acid (5 to 10%). The precipitated material will not pass through the pores of the filters and can be effectively collected on the filter. If the compound of interest has been radiochemically labeled, the filter containing precipitate can be counted by liquid scintillation. Many depth filters commonly serve this purpose, and the collection is generally more rapid than centrifugation.

Materials and Methods

[3]H DNA (human) was a generous gift from D. Kingsbury, University of California, Irvine. Millipore filter, type HAWP (.45 u) was obtained from Millipore Corporation. Whatman No. 1 Filter Paper, Glass Fiber Filter (Reeve Angel Grade 934AH, 2.1 cm), and Sartorius SM111 and SM113 (pore size 0.45 u) filters were obtained from Science Essentials, Fullerton, California. Bovine Serum Albumin (Fraction V) was obtained from Pentex, Inc., Kankakee, Illinois.

---

*   Registered trademark, Millipore Corporation
**   Registered trademark, Sartorius-Membranfilter GMBH
***   Registered trademark, Beckman Instruments, Inc.
****   Registered trademark, Rohm and Haas

Samples were filtered on a standard Millipore filtration apparatus with a fritted glass disc support. Samples were measured in standard glass liquid scintillation vials on a Beckman LS-350 Liquid scintillation counter. The gain was set so that the tritium efficiency in a wide open window was over 60%.

Multichannel analyzer scans were performed on a Nuclear Data Model 1100 (10). All pipetting was done with a Clay Adams dispensing microliter pipette. Cocktail supplies were Beckman liquid scintillation grade.

Results and Discussion

Oftentimes, the type of filter used in a biochemical experiment is determined by cost rather than by performance. Depth filters of the paper or glass fiber type are generally less expensive than that of the membrane type. The coarse nature of the paper and glass fiber type, however, increased the likelihood of beta absorption. Counting efficiencies with these systems are generally less than that of the membrane type.

To compare the efficiencies of various filtering media, [^{3}H]-amino acid serum (195,600 dpm) was placed onto Whatman paper filters and glass fiber filters, allowed to dry for one hour, and counted in three cocktails: toluene/PPO, a non-emulsifying cocktail; Brays solution, a dioxane-based cocktail; and Filter-Solv, an emulsifying and dissolving cocktail. The results, shown on Table I, are compared to the same sample counted in Filter-Solv, but without a filter paper.

Table I

Filter	Cocktail	Toluene/PPO	Filter-Solv	Brays Solution
Whatman Paper	Mean CPM[1]	5083±2.7%	4222±6.2%	3769±1.9%
	ESCR[2]	.775	.720	.725
	SCR[3]	.726	.626	.635
Glass Fiber	Mean CPM[1]	14,223±.7%	22,934±1.4%	12,026±4.0%
	ESCR[2]	.768	.715	.724
	SCR[3]	.712	.672	.626
Without Paper	Mean CPM[1]	–	72,651±1.5%	–
	ESCR[2]	–	.706	–
	SCR[3]	–	.702	–

1. Duplicates were for each cocktail.
2. ESCR based on an LS-350 liquid scintillation counter with a cesium 137 source for ESCR.
3. SCR based on Channel settings of: A=0.300, B=80,300.

A considerable loss in count rate is seen when paper or glass fiber filters are used, mainly due to beta absorption by the intact filter. This causes approximately 95% loss with paper filters and 82% loss with glass fiber filters as compared to the same sample counted without a filter (Table I). None of the cocktails can dissolve these types of filters. Emulsifying type dissolving cocktails (Filter-Solv) can remove some of the material from the glass fibers and suspend it in solution, resulting in a higher counting rate.

Basic Problems in Filter Counting

The measured count rate obtained from a sample on a filter paper can be used to determine the amount of material trapped on the filter. Certain inherent problems with filters in cocktails can affect the measured count rate, irrespective of the actual radioactive material on the filter. This section will illustrate the nature and extent of these problems.

Heterogeneous counting, a term frequently applied to counting a sample on a filter, occurs when the filter remains intact upon exposure to the cocktail. The radioactive sample may thus remain fully imbedded on the filter or be jointly distributed into cocktail and filter. Thus, the radioactive sample can be distributed into two phases and be counted at different efficiencies. Such a situation is commonly encountered with cocktails that have an emulsifier. If this occurs, it may be difficult or impossible to determine the actual effect of quenching in the cocktail.

Heterogeneous counting of samples on filters also poses a further problem in that the filter can absorb some of the beta particles so that the emissions do not produce a measureable excitation of the solvent. This type of quenching is called beta absorption, and can result in a decrease in potential counts from the sample without any direct method of quantitating the degree of quenching. The extent of beta absorption will depend on the energy of the isotope, the type of filter and cocktail used, the degree of heterogeneity in the system, as well as other factors. Whenever a filter remains intact or partially intact in the cocktail, some degree of beta absorption is to be expected.

Intact filters also show geometry and time effects, dependence on the method of application of the sample to the filter, and poor precision. With some filters (usually the screen type), the filters are dried to remove water and make them less opaque than wet filters once cocktail is added. Near transparent filters can often be counted at higher

Table II

Filter Cocktail	Paper	Glass Fibre	Millipore (Mixed Esters)	Cellulose Nitrate	Cellulose Acetate
Toluene (Non-Emulsifier)	H	H	H	H	H
Dioxane	H	H	D	D	D
Dissolving Cocktails (Filter-Solv)	H	H	D	D	T

H - Heterogeneous System; Filter remains intact
D - Dissolved filter
T - Transparent filter

efficiencies than opaque filters, because there is less beta absorption in the system.

Some cocktails are able to chemically dissolve certain filters. While true dissolution of the filter does remove some of the concerns of heterogeneous counting, the aqueous sample must now be rendered into a stable emulsion. Filter-Solv* is a typical example of this class of special cocktails.

Table II summarizes some of the characteristics which arise with common filters and cocktail formulations.

This report will comment extensively on the use of dissolving cocktails such as Filter-Solv*. While chemical dissolution of the filter alleviates problems with heterogeneous counting, the sample must still be placed into intimate physical contact with the solvent. Dioxane-based systems (Brays solution) can dissolve some membrane-type filters, but biological precipitates are not emulsified and heterogeneity can still occur. It is imperative that the filter be dissolved and that the sample be finely emulsified in order to obtain valid counting results, 125-I labeled thyroid stimulating hormone (TSH) was precipitated onto Millipore (HAWP) filters and counted in toluene/PPO where the filter was intact, and in dissolving cocktails (Filter-Solv and Brays). Results are summarized in Table III.

The count rate is higher in Filter-Solv cocktail than in Brays, although both filters are dissolved. While precipitate is finely emulsified in Filter-Solv, the sample (not the filter) precipitates out of solution in the dioxane

* Trademark, Beckman Instruments, Inc.

Table III
Millipore Filters

Cocktail	Mean CPM[1]	ESCR	SCR[2]
Toluene : PPO	3574 ± 3.9%	.671	.846
Filter-Solv	7652 ± 5.7%	.592	.812
Brays	3902 ± 3.3%	.599	.847

1. Values based on duplicate values
2. SCR based on channel settings of: A = 0.300,
   B - 80 300.

based Brays. Therefore, heterogeneity still occurs, and the count rate is lower due to beta absorption. In fact, the amount of beta absorption is comparable to the sample counted in toluene/PPO, where the filter is intact.

Geometry and Position Effects

The actual counting rate of the sample will depend on the position or orientation of the filter in the vial. To demonstrate the effect of positioning, ^{3}H DNA (human) was TCA precipitated onto cellulose acetate membrane filters. The filters were counted in a toluene/PPO cocktail in two positions--0° and 180°. 0° was taken to be a filter at the bottom of the vial with the sample face up; 180° for the sample face down. Filters were either measured immediately when wet or were dried for 1 hour. Dried filters were visually less opaque than wet ones. The counting rates, as a function of position, are shown in Table IV.

A companion set of filters was counted in a dissolving cocktail (Filter-Solv). A higher count rate was seen in this system as the cocktail penetrated the filter further than with toluene/PPO system to increase the contact between radioactive sample and cocktail solvent.

Drying the membrane filters to make them less opaque upon addition of cocktail greatly alleviates the geometry effect. Dry filters gave essentially identical results whether counted at 0° or 180°. Opaque filters, however, generate considerable geometry effects. The measured count rate would decrease by nearly 40% (Table IV) without any significant change in the External Standard Channels Ratio (ESCR).

In the case of opaque filters, a change in the SCR was artificially induced upon measuring the 180° oriented filter. This drop in SCR occurred solely by positioning the filter at

Table IV
Geometry Effects with Cellulose Acetate Filters

Appearance	Position	CPM	ESCR	SCR[1]
Opaque	0°	9,818	.668	.617
(Toluene/PPO)	180°	6,545	.661	.465
Transparent	0°	26,574	.650	.626
(Toluene/PPO)	180°	26,542	.645	.609
Transparent	–	29,651.6±2.9%[2]	.595	.543
(Filter-Solv)				

1. SCR based on channel settings of: A = 0.300, B = 80.300
2. Duplicate average

180° - there was no increase in quenching in the system. The ESCR also does not appreciably change to reflect the loss in counting efficiency due to positioning of the filter.

There is nearly a 12% CPM difference between filters that are effectively "transparent" or dried and those counted in a suitable dissolving cocktail.

In a separate experiment, another batch of [^{3}H] DNA (human) was TCA precipitated onto cellulose nitrate membrane filters (Sartorius-SM113). Similar effects were found which are summarized in Table V.

As in the case of cellulose acetate membrane filters, the cellulose nitrate filters show that positioning the filter affects the observed counting rate. There is

Table V

Geometry Effects with Cellulose Nitrate Filters

Appearance	Position	CPM	ESCR	SCR[1]
Opaque	0°	20,276	.672	.622
(Toluene/PPO)	180°	13,247	.696	.486
Transparent	0°	25,685	.681	.592
(Toluene/PPO)	180°	25,869	.669	.580
Dissolved	–	36.831±1.1%[2]	.584	.632
(Filter-Solv)				

1. SCR based on channel settings of: A = 0.300, B = 80.300
2. Duplicate Average

approximately a 40% increase in count rate between "transparent" cellulose nitrate filters and dissolved filters. This suggests that filters which are transparent or near transparent to the naked eye can still experience extensive quenching by beta absorption.

Changing counting rates

Opaque filters will commonly show a changing count rate for the same sample with time. In some cases where an emulsifying cocktail is used, material slowly elutes off the filter and is emulsified into the cocktail solution. The emulsified material is in better physical contact with the cocktail and will be counted at a higher efficiency than that which remains on the filter; hence, the count rate will rise with time.

A changing count rate has also been found with non-emulsifying cocktails (toluene/PPO). In these cases, the rise in counts stems from a progressive penetration of the dried filter by the cocktail--better penetration leads to better physical contact between the sample and cocktail, and a higher counting rate results.

To illustrate this, radioactive material in the form of [^{3}H] DNA (human) was filtered on cellulose acetate membrane filters (Sartorius SM 111) and positioned at 0° in a vial. The samples were counted for various time intervals in a standard scintillation vial with 10.0 ml of non-emulsifying toluene/PPO cocktail. Two counting channels were used--the first channel covering the entire tritium energy range, and the second channel covering the upper energy tritium events. A ratio of the counting rate in the second channel divided by the first is a typical form for the SCR method. The counting rates of the filter sample as a function of time of exposure to cocktail are shown in Table VI.

While the count rate continues to increase as the cocktail continues to penetrate the filter, the ESCR does not change. The SCR, however, does undergo an initial change by decreasing .041 units to erroneously suggest a decrease in counting efficiency even though the count rate has increased.

The initial decrease in SCR stems from an increased sensitivity to low energy tritium decays as the cocktail penetrates the filter. This increased penetration of cocktail improves the intimate contact between sample and cocktail solvent so that weak beta decays which were previously absorbed can now produce a measurable pulse of light. Essentially, all of these pulses are from low energy decays and the second counting channel (which forms the numerator of the SCR) does not count as many of these low

Table VI

Time of Counting (hours)	Gross CPM Channel A	Gross CPM Channel B	ESCR	SCR[1]
0	2409	1608	.669	.668
1	3351	2101	.665	.627
2	5706	3491	.666	.612
3	7212	4473	.668	.620
24	9818	6057	.668	.617

1. SCR based on channel settings of: A = 0.300, B = 80.300.

energy decays as the first channel. After 1 hour, this penetration stabilizes. It is important to mention that these weak beta decays were occurring in the sample as soon as cocktail was added--they were just absorbed by the filter to prevent the production of any light pulse.

The initial change in SCR is, of course, not due to any genuine change in chemical or color quenching. In fact, the ESCR which effectively measures the cocktail solution for quench content is the same for all measurements. After low energy penetration by the cocktail, further penetration produces a uniform increase in pulses throughout the entire LS spectrum to render the SCR unchanged. The count rate, however, increases accordingly.

In Table VII, results are shown which demonstrate the time effects with cellulose nitrate membrane filters.

Table VII

Time Effects

CPM[1]

Time	Channel A	Channel B	ESCR	SCR[2]
0	8,930	6,854	.785	.768
30 min	10,757	7,982	.782	.742
60 min	11,252	8,292	.786	.734
90 min	12,049	8,666	.783	.719

1. Sample applied to cellulose nitrate filters and counted in a toluene/PPO cocktail.
2. SCR based on channel settings of: A = 0.300, B = 80.300.

Table VIII

	Application	
	Spread Over Filter	Topically Applied
Mean CPM on Filter[1]	13,400 ± 642	4,131 ± 821
Mean CPM of Effluent	138 ± 31	206 ± 10
Mean ESCR	.767 ± .001	.768 ± .001
Mean SCR[2]	.709 ± .010	.702 ± .004

1. Based on average of five filters
2. SCR based on channel settings of : A=0.300, B=80.300.

Dry filters do not show such a strong dependence on time of exposure to cocktail. Apparently, the removal of water from the filter facilitates a rapid penetration of filter by the cocktail, and a stable count rate is obtained.

Application of Sample to Filter

The counting rate of samples on filters will be affected by the surface area of exposed sample. Samples which are spread over the entire filter have a greater exposed surface area to the cocktail than samples which are applied topically.

A set of seven identical samples containing the same amount of [3H] DNA (human) were precipitated and applied onto Millipore filters. Four were finely spread over the entire surface of the filter, while the remaining three were topically applied onto a small surface of the filter. All samples were dried and made transparent in a toluene/PPO cocktail. The counting rates of the two samples are shown in Table VIII.

High coefficients of variation on filters

Intact filters typically show a high coefficient of variation for replicates. Samples counted on a filter paper often show count rates which depend greatly on the geometry of counting, the method of application, the degree of dryness, and other physical parameters. All of these parameters contribute to produce a coefficient of variation outside of the expected experimental deviation of pipetting and counting statistics.

An example of the high coefficient of variation that is encountered with typical filters is shown in Table IX. In this example, 5 filters were prepared by TCA precipitating

Table IX

Cocktail	Mean CPM[4]	Mean ESCR	Mean SCR[5]
Toluene: PPO, wet[1]	6,671.9±39%	.777±9.2%	.711±1.3%
Toluene: PPO, dry[1]	27,925.6±11%	.774±1.6%	.719±1.0%
Filter-Solv[2]	39,056.1±2.8%	.706±1.0%	.692±.8%

1. Millipore filters, 5 replicates.
2. Cellulose nitrate filter, triplets.
3. Coefficient of variation (percent).
4. All samples counted to a 2% error.
5. SCR based on channel settings of: $A = 0.300$, $B = 80.300$.

tritiated DNA onto standard cellulose nitrate filters. The filters were collected and counted in toluene/PPO. A second set of 5 filters was prepared in the same way but was dried for 1 hour before adding cocktail. The physical parameters that affect the count rate exert a strong effect on wet filters. Drying the filters does reduce this coefficient of variation, but it is nevertheless quite high. As a baseline for comparison, identical filters were dissolved in a special dissolving cocktail (Filter-Solv) and counted.

Beta Absorption on Filters

Beta absorption occurs when filters remain intact, or partially intact upon addition of cocktail. Beta emissions can occur within the matrix of the filter and sample without exciting the cocktail to produce light scintillations. When an intact filter is counted, the observed count rate is due to radioactive decays that actually leave the filter paper and travel into the cocktail.

The degree of absorption is, in part, related to the amount of material on the filter--the more material on the filter, the greater the probability that some of the decays will be absorbed. It is common practice to add an unlabeled carrier substance to the radioactive material, and precipitate the entire mixture with an acid solution (trichloro-acetic acid is commonly used). The carrier facilitates the trapping of the radioactive material on the filter by coprecipitation. If an insufficient amount of carrier is used, however, the radioactive material will pass through the filter and not be counted. If too much carrier is used, the degree of absorption increases.

To illustrate in a typical application, tritiated DNA (Human) was TCA precipitated on a standard Millipore filter

(type HAWP). Varying amounts of an unlabeled carrier
(Bovine Serum Albumin) were used to facilitate trapping the
DNA on the filter. After drying the filters for over 60
minutes, duplicates were counted in a toluene/PPO system
with an ESCR (External Standard Channels Ratio) as well as
an SCR (Sample Channels Ratio) taken for each sample. Upon
addition of cocktail to the dried filters, they became
relatively transparent to the naked eye.

For comparison, an identical series was prepared, but
the samples were dissolved in a special dissolving cocktail
(Filter-Solv). The 50 microliter aliquot of tritiated DNA
was also counted directly in a cocktail without the use of a
filter. Its count rate was approximately 44,000 counts per
minute. Results are summarized in Table X.

Note, that with relatively small amounts of carrier
(under 200 micrograms) a certain amount of the DNA will not
be trapped on the filter. Filters dissolved without any
carrier, have nearly the same count rate as the dried filters, but nearly 14,000 of the 44,000 total counts per
minute in the DNA aliquot are lost through the filter. If
carrier is included with the precipitation, the count rate
increases as more radioactive material is trapped on the
filter. A plateau of nearly 40,000 counts per minute is
trapped on the filter from 200 to 700 micrograms of carrier.
This is not seen, however, in filters that are counted
heterogeneously (transparent but undissolved) as the level
of beta absorption increases as the amount of unlabeled
carrier is increased. With this type of DNA and this type
of Millipore filter, the plateau is not reached until 200
micrograms of carrier is used. The filter really contains
approximately 40,000 counts per minute of DNA, but the
highest count rate with transparent filters is less than
25% of that of the dissolved filter. Over 75% of the
material on the 200 microgram BSA filter has been absorbed.
Moreover, this level of absorption increases as more carrier
is used.

Conventional instrumental methods of assessing quenching (the loss in the counting efficiency) do not reflect the
reduction in counts due to beta absorption. While the ESCR
(External Standard Channels Ratio) and the SCR (Sample
Channels Ratio) between the transparent filters and the dissolved filters cannot be compared directly, since different
cocktails are used, the ESCR and the SCR for all transparent
filters remain relatively unchanged even though the count
rate has decreased from 28,320 to 3,351 counts per minute.

The ESCR is not expected to change, as the loss in
count rate stems from the heterogeneous distribution of

Table X
Effect of Carrier on Absorption of Sample

Appearance of Sample	Micrograms BSA	CPM	ESCR	SCR[1]
Transparent Filter[2]	0	28,320±1.149	.776	.731
Dissolved Filter	0	29,950	.715	.706
Transparent Filter	100	22,832±50	.774	.739±.001
Dissolved Filter	100	34,433	.704	.670
Transparent Filter	200	9,656±316	.772	.744±.002
Dissolved Filter	200	39,449	.708	.691
Transparent Filter	300	7,247±742	.775±.001	.736±.005
Dissolved Filter	300	39,157	.709	.698
Transparent Filter	500	4,781±664	.773±.001	.733±.005
Dissolved Filter	500	37,947	.697	.680
Transparent Filter	700	3,351±469	.773±.001	.733±.001
Dissolved Filter	700	39,705	.703	.694

1. SCR based on channel settings of: $A = 0.300$, $B = 80.300$.
2. Duplicates were made for each set.

radioactive material in a non-scintillating phase (the
matrix of the filter). In the samples where the filter is
dissolved, the ESCR values are lower than the ESCR values
of the transparent filters because some chemical quenching
has been introduced into the system through the dissolution
of the filter and the emulsification of the sample.
     The SCR is based on the energy distribution of the
tritiated sample. If chemical or color quenching is present
in the sample, the resulting distribution of the tritiated
sample will change. Two preselected counting channels of
the LS counter, which are set to measure discrete portions

of the sample's LS spectrum, will show a change when the two
channels are expressed as a dimensionless ratio. When quenching is present, the counting rate will change, but the distribution of the LS spectrum will also change and the SCR
will decrease to reflect the effect of quench. In the case
of the filter, however, the SCR does not change in accordance
to the increase in absorption.

To demonstrate this, after counting the samples on an LS
counter, the samples were placed in a multichannel analyzer
which effectively counts the samples for time increments over
small energy ranges to obtain an LS spectrum of the radioactive sample.

As more carrier is added to the filters, the absorption
increases and the counting rate decreases accordingly. The
distribution of the sample's LS spectrum, however, does not
change, and therefore, the SCR for these samples does not
change either. Figure 1 shows the multichannel scans of
these filtered samples. Increasing beta absorption merely
causes a diminution in the total LS spectrum; the distribution is not affected, and the SCR, therefore, does not
change. Figure 2 shows a comparison of a transparent filter
and an identical aliquot in a dissolved filter. The difference in area between the two curves is due to the absorption of the radioactive material on the transparent filter.

In general, heterogeneity produces beta absorption
which can lower the potential count rate for a sample on a
filter without changing the ESCR or the SCR. In addition to
inducing heterogeneity in a system by the sample carrier
(this type of beta absorption is more appropriately called
"self absorption"), heterogeneity can originate from other
ways.

Wet filters will have more absorption than filters
which are dried. Heterogeneity is also a function of the
length of time a filtered sample is allowed to dry. Cellulose nitrate filters (Sartorius SM 113) were used to trap
TCA precipitated DNA (tritiated; approximately 44,000 counts
per minute applied to each filter) and each filter was dried
for a different length of time before counting in a toluene/
PPO cocktail. To determine the true amount of activity
trapped on the filter, a similarly prepared filter was
counted in a special dissolving cocktail (Filter-Solv). Both
the ESCR and the SCR was taken for the filters and the
results are shown in Table XI.

As the filters are dried for longer times, the counting
rate rises. Both the ESCR and the SCR, however, remain
unchanged. As the filters are dried longer, less water
remains on the filter to inhibit cocktail contact with the

CARRIER EFFECTS

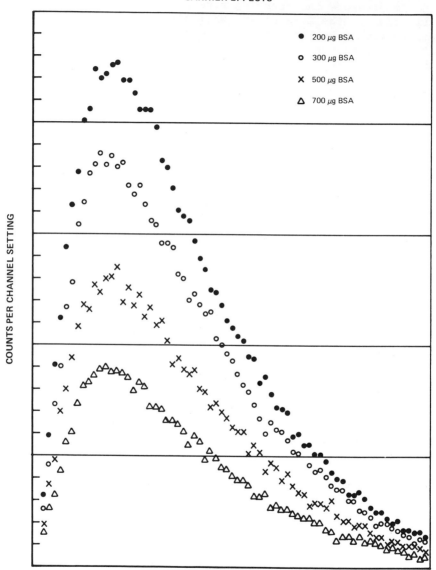

Carrier Effects on DNA Samples

FIGURE 1

CARRIER EFFECTS

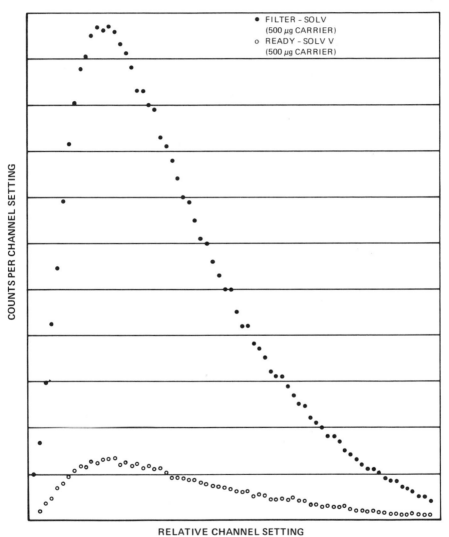

Comparison of Absorption Effects on Dissolved Versus
Undissolved Filter
FIGURE 2

Table XI

Sample	Drying Time	Appearance	CPM	ESCR	SCR[1]
D	0	Opaque	12,325	.782	.703
A	15 min	Opaque	16,880	.777	.716
B	60 min	Opaque	26,539	.778	.705
C	1 day	Transparent	32,785	.773	.711
Filter-Solv	–	Dissolved	36,881	.697	.692

1. SCR based on channel settings of: A = 0.300, B = 80.300.

sample. The improved penetration of cocktail on the filter reduces the amount of beta absorption on the filter. As beta absorption merely affects the total LS spectrum of the sample and not the energy distribution, the ESCR and the SCR should not change to reflect the improved counting efficiency of the dried samples. At least 36,000 counts per minute were present on each filter, but none of the intact filters indicate the true activity on the filters. Even the filter dried for 1 day experiences some degree of beta absorption.

Multichannel scans were taken of the samples and their spectra are shown in Figure 3. The decrease in absorption results in an increase in counting rate, but the SCR and ESCR fail to show this effect.

As long as absorption can occur, the true activity on the filter can be "masked" or hidden, and interpretation of results based solely on the count rate with filter heterogeneous systems can lead to many erroneous conclusions.

Consider a series of tritiated DNA samples of varying activity which was TCA precipitated in the presence of different amounts of carrier protein (Bovine Serum Albumin). The precipitates were collected on Millipore filters (HAWP) and dried for over one hour to increase their transparency once cocktail (toluene/PPO) was added. Companion sets were counted in a dissolving cocktail (Filter-Solv) and the results are shown in Table XII. The actual counts per minute applied to the filter was determined by adding the tritiated aliquots directly without any filtration to an emulsifier cocktail. The filtered effluents were also counted to ensure that the material was not passed through the filter. The average activity of the effluents was determined to be approximately 303 cpm.

While some filters have up to 100% more radioactive material on them than others, the count rates with the heterogeneous systems are clearly unreliable. The ESCR and

CARRIER EFFECTS

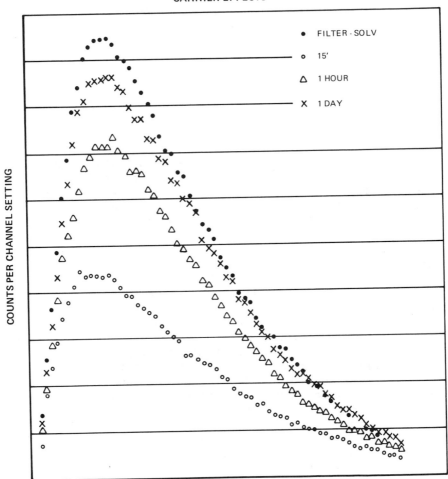

Drying Time Effects Versus Dissolved Filter

FIGURE 3

Table XII

Actual CPM Applied	µg BSA	Mean[1] CPM of Filter	Mean ESCR	Mean SCR[2]	Mean CPM Dissolved Filter
41,191	200	7667±552 (374)[3]	.771±.001	.717±.007	39,842±748
63,653	300	8833±437 (458)	.771±.002	.722±.008	59,623±738
82,544	500	8696±261 (178)	.769±.001	.721±.001	80,547±1382

1. Based on duplicate samples.
2. SCR based on settings of: A = 0.300, B = 80.300.
3. CPM of wash.

the SCR remain unchanged because the count rate on the filters depends solely on the extent of beta absorption. Previous experiments have shown that beta absorption can greatly affect the counting rate without producing any change in the Sample Channels Ratio or the External Standard Channels Ratio.

Dissolved filters can be expected to give a true reflection of the actual material on the filter. They all experience an average loss of about 4% of the total activity (CPM) applied to the filter. Several factors such as chemical quenching caused by dissolution of the filter, and loss of material through the filter contribute to this 4% reduction in the counting rate.

This is not a 4% loss in true activity in the sense of disintegrations per minute (DPM), but rather a loss in maximal counting rate (CPM). In many biochemical experiments (such as labeling DNA), it is not possible to know the actual DPM due to un-ertainties in the labeling process and all values are expressed as a "relative counting efficiency". On this basis, the filters dissolved in Filter-Solv (ESCR of .696, SCR of .686) give a 96% relative counting efficiency and the actual corrected count rate can be obtained by dividing the observed count rate by .96.

Conclusions

The examples discussed in this report illustrate many of the difficulties in counting radioactive material on filters. The observed count rate will depend on a number of physical parameters in the preparation of the samples for counting. Moreover, the heterogeneous nature of the sample counted will cause absorption of the radioactive emissions. The absorption seen on many of the samples occurred with filters which were effectively "transparent". Due to variations in sample handling, it may be impossible to determine the extent of absorption on each sample that is filtered.

The cocktail used for these filter samples cannot be lightly regarded. Toluene/PPO has many attractive advantages. Its ease in preparation and inexpensive cost lends itself well to counting filters. The use of xylene or other fluors may improve the scintillation efficiency of the cocktail somewhat, but in general, these non-emulsifying cocktails will result in a heterogeneous sample to count. The absorption generated by these conditions can cause the count rate of the sample to be unindicative of the true activity on the filter.

Drying the screen type filters (membrane) before the addition of the cocktail is recommended, but this procedure is time consuming and may not always alleviate all of the difficulties. Geometry effects are, of course, lessened as the dry filters are less opaque than moist ones, but the counting rate is greatly dependent on the surface area of exposure. Absorption is still present on filters even after extended drying times.

Many common commercial emulsifying systems, however, are reported to produce heterogeneity with depth or screen filters (11). Frequently, the cocktail that dissolves the filter will be counted at an apparently lower efficiency than with a heterogeneous system. The organic based components which can dissolve a cellulose polymer will usually be unsuitable scintillation solvents and the ESCR and SCR for such samples are generally lower than with heterogeneous samples. Unlike heterogeneous systems, however, the ESCR and the SCR of dissolving cocktails truly reflects the quenching that has occurred in the system.

By appropriate use of standards, the actual activity on the filter can be determined by an appropriate "correction factor". This correction factor is slightly different than conventional chemical or color quenched corrections as a finite amount of material can pass through the filter. The loss in count rate due to amount of material loss through the filter (assumed to be constant for given type of filter under a given set of conditions) can be included with the conventional loss in count rate due to quenching to produce this correction factor.

Based on studies of sulfur-35, Bush has suggested the use of a "double ratio" of the ESCR and the SCR to indicate

67

heterogeneity in samples. The tacit assumption in this
technique is that the ratio of the SCR divided by the ESCR
changes with absorption because beta absorption will affect
the beta spectrum (12). The method was originally intended
to be a qualitative indication of sample heterogeneity.
Certainly, heterogeneous samples can show changes in their
SCR values, but such a generalization is not always valid.

It appears that the weak isotopes, such as tritium, do
not always show a change in their beta spectrum. In the
examples cited in this report, neither the SCR or the ESCR
show the extent of absorption or heterogeneity, and the value
of a double ratio for these samples is limited. With the
weak beta emitters, the count rate depends on the material
in surface contact with the cocktail and the beta spectrum
cannot always reflect the absorption within the filter.

Acknowledgements

The authors would like to thank Beckman Instruments for
its support of this work. We would also like to thank Dr. D.
Kingsbury of the University of California at Irvine for his
donation of samples and many useful discussions. We would
especially like to thank Dr. Donald Horrocks for his insight
and help throughout this work.

Bibliography
1. "DNA Synthesis In Vitro" (Well, R.D. and Inman, R.B. ed.)
   University Park Press, Baltimore (1973).
2. J.W. Geiger and L.B. Wright, Biochem. Biophys. Res.
   Comm. 2, 282 (1960).
3. R.B. Loftfield, Atomlight, No. 13, New England Nuclear
   Corp., Boston Mass. (1960).
4. J.D. Davidson, Proc. Conf. Organic Scintillation Detectors, Univ. of New Mexico, 1960 (G.H. Daub, F.N. Hayes,
   and E. Sullivan, eds.), TID-7612, p. 232, U.S. At.
   Energy Commission, Washington, D.C. (1961).
5. N.B. Furlong in The Current Status of Liquid Scintillation Counting (E.D. Bransome Jr., ed.), p. 201, Grune
   and Stratton, New York, 1970.
6. E.D. Bransome Jr. and M.F. Grower, Analytical Biochemistry 38, 401 (1970).
7. R.E. Johnsonbaugh,Analytical Biochemistry 54, 490-4 (1973).
8. R.M. McKenzie, Analytical Biochemistry 54, 17-31 (1973).
9. A. Chakravarti, Analytical Biochemistry 40, 484 (1971).
10. D.L. Horrocks, Nucl. Instr. Meth. 117, 589 (1974).
11. Multiple Sample Filtration and Scintillation Counting,
    Millipore AB 304, p. 12.
12. E.T. Bush, Internat. J. of Applied Radiation and Iso-
    topes 19, 447 (1968).

LIQUID SCINTILLATION IN MEDICAL DIAGNOSIS

Kent Painter
Department of Radiology and Radiation Biology
Colorado State University
and
Micromedic Diagnostics, Inc.
Fort Collins, Colorado

With the tremendous increase in the application of radioassay, particularly radioimmunoassay, in the clinical laboratory liquid scintillation counting became an indispensable
tool in diagnostic medicine. Few publications, however, have
concerned themselves with problem areas which occur with the
method in the clinical laboratory. The purpose of this presentation is to summarize our experiences with the liquid
scintillation technique in the clinical situation.

The Digoxin Example

Before entering into a discussion on the application of
radioassay in the clinical laboratory, it is necessary to
inspect data from a typical radioassay. In order to appreciate the relative importance of a diagnosis in this area,
we have selected the radioimmunoassay of digoxin as a particular example.

Digoxin is a common drug of the digatalis or cardiac glyco-
side group used to treat advanced heart disease in humans.
Due to a number of patient variables the exact doseage of
digoxin must be tailored to the individual and radioimmunoassay is used to monitor serum levels. Too low a doseage
of digoxin renders the drug ineffective or non-therapeutic.
Too high a doseage has been implicated in the deaths of
patients due to severe toxicity above the therapeutic level
(1-3). The doseage for each patient must be carefully
adjusted to maintain serum levels in a very narrow therapeutic range of about 1-2 ng/ml.

Table 1 presents data obtained from a typical digoxin radioimmunoassay using a tritiated digoxin derivative. In order
to achieve the requisite sensitivity for the assay a
limited amount of tracer mass is added to each assay tube.
Thus, the subsequent counting rate is quite low within the
range of diagnostic significance. It is easy to see, therefore, that a small error in the counting rate will have a

significant effect on the ultimate diagnosis.

Note that the background counting rate is almost identical to the counting rate of the last sample. This means that a long counting time must be employed to ensure accurate measurement of both background and sample counting rates. Secondly, notice that the therapeutic range, that is between 1-2 ng/ml, comprises a narrow counting rate range of only~200 cpm and a small error in measurement of the counting rate could result in a faulty diagnosis. Therefore, the effect of various sources of background or spurious counts becomes quite important in liquid scintillation digoxin radioimmunoassay.

Chemiluminescence

Many clinical users of the liquid scintillation technique are totally unaware of the possibility or consequences of chemiluminescence. To demonstrate the effect chemiluminescence can have on an assay, 300 ul of a simulated non-radioactive digoxin radioimmunoassay sample were added to 3.0 ml of a Triton:toluene (1:2) scintillation cocktail (4). The samples were placed in a liquid scintillation counter and counted for one minute repeatedly for five hours. Even at 130 minutes after insertion of the sample into the liquid scintillation counter (Figure 1), the rate of chemiluminescence was still three times background. Since the assay for digoxin in many cases needs to be a stat procedure, one wonders whether the liquid scintillation technique should be used. Even in the best commercial liquid scintillation cocktails, which have reducing agents and acidic buffers to minimize chemiluminescence, the rate of spurious counts was still six times background at one hour after preparation of the sample.

Table 2 demonstrates that as expected the chemiluminescence counting rate increases dramatically as the serum sample size increases. Therefore, the smallest practical sample should be used.

It is imperative that samples be allowed to equilibrate preferably for four hours in the liquid scintillation counter prior to counting. The use of a chemiluminescence monitoring device, such as the Photon Monitor, is highly recommended in these cases. However, these devices are set at an arbitrary luminescence rate which may be unsatisfactory in certain assays. Users of this type of instrumental rejection of samples should consult their manual to be certain that the

Table 1. Typical Tritiated Digoxin Radioimmunoassay[1]

Sample CPM	Net CPM	Digoxin Concentration ng/ml	
1245	1061	0	Non-therapeutic
1137	953	0.4	
962	778	1.0	Therapeutic
746	562	2.0	
628	444	3.0	
502	318	5.0	Toxic
372	188	10.0	

Background - 184 cpm

[1] Schwarz/Mann Digoxin Radioimmunoassay Kit brochure (undated).

Table 2.

The Effect of Sample Volume on Chemiluminescence Rate

Decreasing volumes of a non-radioactive serum were added to 3.0 ml cocktail in a mini-vial. The chemiluminescence rate was measured for one minute five minutes after mixing.

Sample Volume (microliters)	Chemiluminescence Rate (Net cpm)
500	4742
400	2110
300	1002
200	493
100	147

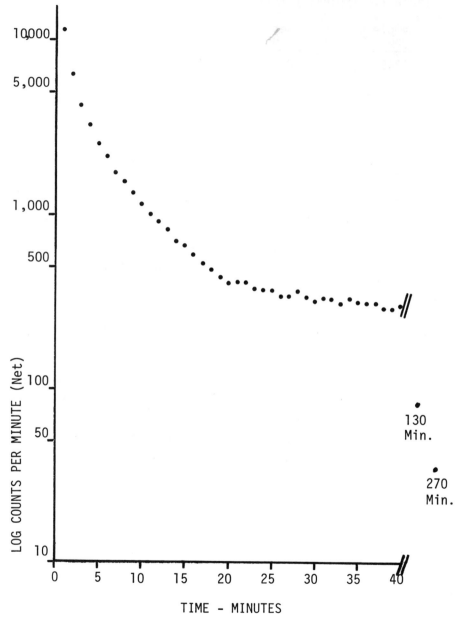

Figure 1   CHEMILUMINESCENCE OF RADIOIMMUNOASSAY SAMPLE

rejection rate is low enough for the particular assay in question.

Photoluminescence is caused by exposure of vials, caps and cocktail to direct sunlight or fluorescent lighting. Black caps are less prone to photoluminescence than white caps (5). Cocktail solution should be stored in amber bottles in the dark and should be allowed to stabilize for 24 hours after preparation. Direct sunlight and fluorescent lighting should be excluded from the counting laboratory.

Refering back to our digoxin example in Table 1, an unde- tected increase in background or chemiluminescence will re- sult in a falsely low measure of digoxin. This may lead to the prescription of an increased doseage to the patient with the resultant risk of digitoxicosis.

Counter Performance

While the quality control of radioimmunoassays is usually quite rigid in terms of the characteristics of the assay calibration curve, one must also monitor the characteristics and shape of the quench correction curve. Figure 2 is a photograph of our counter quality control sheet which dis- plays a problem occurring with one of our spectrometers when it began to experience phototube fatigue. One can see from the dates that over the period of one month the sample channels ratio calibration curve drifted considerably out of the range normally expected. This resulted in inaccurate quench correction factors for the assay. When the faulty phototube was replaced, the quench correction curve returned to normal and remained stable. It is recommended that the quench correction curve be monitored and recorded with each run as a means of measuring instrument performance.

Quench Correction

It is not at all uncommon for commercial radioimmunoassays or publications on radioimmunoassay to present data in cpm which has not been corrected for quenching. While most of us in this audience are well aware of the dangers in attempting to tabulate data which has not been corrected for quenching, it is common practice, I am afraid, in many clinical laborato- ries.

In order to demonstrate the effect of both chemical and color quenching of different sera, 8 human and 2 commercial control

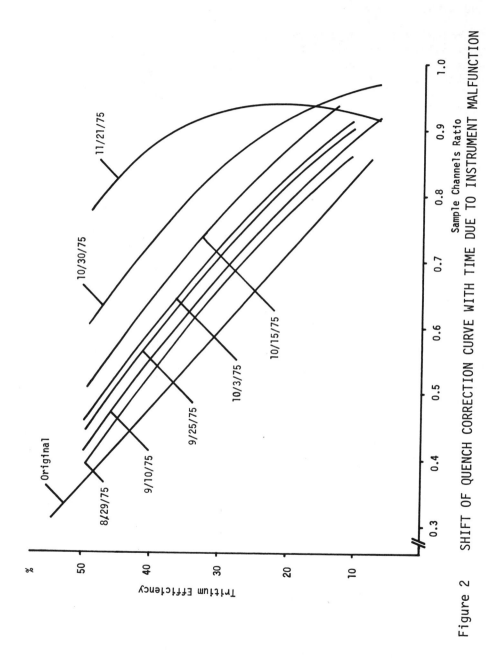

Figure 2   SHIFT OF QUENCH CORRECTION CURVE WITH TIME DUE TO INSTRUMENT MALFUNCTION

74

sera were analyzed by the manner described above. In this case, however, the liquid scintillation cocktail was spiked with 30,120 cpm ± 401 cpm ($\overline{X}$ ± σ) tritiated toluene and all vials were precounted prior to the addition of serum. Samples were equilibrated 24 hours in the dark to eliminate chemiluminescence.

Table 3 indicates the loss in counts due to quenching of different sera. Based upon the typical digoxin counting data in Table 1 one can see that a significant error is introduced when quench correction is not employed. The range of counting rates in the samples with identical amounts of tritium was 3073-6058 cpm. It is also interesting to note that commercial control sera, which are usually charcoal stripped human sera, do not count with the same relative counting efficiency as patient sera.

Color quenching can be reduced and counting efficiency can be improved by using a smaller sample and by maximum dilution of sample in scintillation fluid. The use of 20 ml vials diluted to the very top helps to minimize the effect of color quenching. When highly colored sera are encountered in a digoxin assay, it is mandatory that internal standardization techniques by employed. Since it is well known that external or sample channels ratio methods of quench correction do not totally correct for color quenching, the counting rate of the sample will be grossly underestimated. In the case of the digoxin radioimmunoassay the patient's serum would appear to contain more digoxin than is actually present. Based upon the assay alone a physician might prescribe a decreased doseage of digoxin.

Contamination From In Vivo Scanning Agents

Although it is rare for a contaminated radioactive serum sample to come into the laboratory for a radioimmunoassay test, a number of documented cases have occurred and one must be prepared for this eventuality. One must ensure that the serum sample has not been taken from the patient following a nuclear medicine in vivo scanning test. Since these agents are invariably strong gamma emitting nuclides, they can be picked up rather readily in most cases by looking at an auxiliary channel which has a lower discriminator set beyond the tritium beta spectrum end point. A quick glance at the counting data from the auxiliary channel will reveal radioactively contaminated sera.

Table 3   Quenching Effect of Different Sera

Cocktail was pre-spiked and each vial pre-counted to
34,120 ± 401 dpm prior to the addition of sample.

Human Sera	Net Counts per Minute
A	4308
B	5100
C	3082
D	3565
E	5548
F	5130
G	3769
H	3073
Commercial Control Sera	
A	5957
B	6058

Sample Homogeneity

Many liquid scintillation cocktails will not adequately dissolve the strong buffers and large amounts of serum used in radioummunoassay. Samples should be inspected visually after the four hour equilibration period, particularly when refrigerated counters are used. A sample which is apparently homogeneous at ambient temperature may undergo phase separation at 8°C.

The usual way to monitor sample homogeneity is to utilize the double ratio test of Bush (6). This method involves the use of both sample channels ratio and external standard channels ratio to compute counting efficiency. A sample which results in grossly different efficiency determinations by both sample channels ratio and external standard channels ratio is nearly always heterogeneous. Several computer programs have been developed to test each sample by the double ratio method and to flag aberrant samples. A good reference on this topic is Glass (7).

Recommendations

Based upon our experience with clinical samples we offer the following suggestions which are intended to minimize spurious results.

1. Utilize 20 ml glass counting vials with black caps (5).
2. Utilize the minimum practical sample size.
3. Dilute the sample to the maximum capacity of the counting vial.
4. Add a reducing agent to the cocktail (8).
5. Maintain a pH <7 in the cocktail (9).
6. Adopt the double-ratio test (6).
7. Monitor quench correction curves with each run.
8. Avoid exposure of samples to direct sunlight or fluorescent lighting.
9. Never open the counter lid while samples are counting.
10. Allow samples to equilibrate a minimum of four hours and inspect them prior to counting.
11. Monitor an auxiliary channel for radioactively contaminated samples.
12. Use a Photon Monitor or similar device to detect chemiluminescent samples.
13. For highly colored sera correct for quenching by internal standardization.

Summary

In summary one might conclude that because of the pitfalls which can occur in the liquid scintillation technique, the method has fallen from being a useful analytical tool to somewhat of an unpredictable, unreliable technique in the clinical laboratory. Radioimmunoassay users have gone to great lengths to circumvent liquid scintillation counting by labelling compounds with gamma or X-ray emitters, so that a "gamma" counter can be substituted. In that sense I believe we have failed the medical community by not devising novel, simple solutions to the problems described above. Hopefully, sessions such as these will lead to simplifications and improvements in liquid scintillation technology which will encourage, rather than discourage, its use as an analytical tool to solve problems in medicine and research.

References

1. Beller, G.A., Smith, T.W., Abelmann, W.H., Haber, E. and Hood, W.B., New Eng. J. Med., 284, 989 (1971).
2. Editorial, Medical World News, 12:14 (1971).
3. Holt, D.W. and Benstead, J.G., J. Clin. Path., 28, 438 (1975).
4. Turner, J.C., Int. J. Appl. Radiat. Isotopes, 19, 557 (1968).
5. Painter, Kent and Gezing, M.J., Anal. Biochem. 58, 334 (1974).
6. Bush, E.T., Int. J. Appl. Radiat. Isotopes, 19, 447 (1968).
7. Glass, D.S., in Organic Scintillators and Liquid Scintillation Counting, Horrocks, D.L. and Peng, C.T., eds., Academic Press, New York, p.803.
8. Houtman, A.C., Int. J. Appl. Radiat. Isotopes, 16, 65 (1965).
9. Kalbhen, D.A., in The Current Status of Liquid Scintillation Counting, Bransome, E.D., ed., Grune and Stratton, New York, 1970, p.337.

THEORY AND APPLICATION OF CERENKOV COUNTING

H. H. Ross
Analytical Chemistry Division
Oak Ridge National Laboratory*
Oak Ridge, Tennessee 37830 (USA)

ABSTRACT

The production of Cerenkov radiation by charged particles moving through a transparent medium (the Cerenkov generator) is a strictly physical process that is virtually independent of the chemistry of the medium. Thus, it is possible to calculate easily all of the important parameters of the Cerenkov process such as excitation threshold, emission intensity, spectral distribution, directional characteristics, and time response. The results of many of these physical effects are unknown in conventional liquid scintillation spectroscopy and, therefore, many unique assay techniques have been developed as a result of their consideration. This paper will discuss the theoretical basis of the Cerenkov process, explore the unusual characteristics of the phenomenon, and demonstrate how these unusual characteristics can be used to develop equally unusual counting methodologies. Particular emphasis will be directed toward an analysis of published data (obtained with conventional liquid scintillation instrumentation) that would be difficult or impossible to obtain with other counting techniques. Also discussed is a new type of sample vial that uses an isolated waveshifting system. The device was specifically designed for Cerenkov counting.

---

*Operated by the Union Carbide Corporation for the Energy Research and Development Administration.

> Admiring, in the gloomy shade,
> Those little drops of light.

Edmund Waller

INTRODUCTION

Although the concept of counting a radionuclide dissolved in a liquid <u>via</u> Cerenkov radiation was demonstrated by Belcher (1) over 20 years ago, it is only within the last five years that a significant number of real applications have appeared in the open literature. This lack of interest was most probably due to the difficulty in measuring the very feeble emission generated by beta particles in the range of 0.5 to 2.0 MeV. For example, a 1.0 MeV beta particle in water dissipates only about 1 KeV of energy in the 300 to 700 nm spectral region (perhaps 30-45 photons). Add to this tenuous photon emission the limited spectral sensitivity and conversion efficiency of early photo-multiplier tubes, and the experimental difficulties become obvious. Other problems, such as dark current (S/N), directional considerations, and noisy electronics only compounded the dilemma. However, new instrumentation concepts and components for liquid scintillation counting have changed the situation significantly; high sensitivity Cerenkov counting has become a routine operation in many laboratories.

Jelley (2) has written an excellent account of the theory and application of Cerenkov radiation. The major application and instrumentation emphasis is, however, directed toward problem areas in high energy physics. More recent discussions (3-5) cover the general area of using commercially available liquid scintillation spectrometers for the assay of dissolved nuclides in small, conventional samples. These papers describe a wide range of experimental conditions that include typical detection efficiencies, instrument set-up, quenching effects, threshold energy response, sample volume, and directional considerations. All of these papers emphasize three important considerations peculiar to Cerenkov counting - energy discrimination, chemical quenching, and the absence of conventional scintillation fluors in the system. A view of these special characteristics reveals the unique application areas of Cerenkov spectroscopy.

Energy discrimination

Perhaps the most unusual aspect of Cerenkov counting is the presence of an energy threshold effect that is solely a function of the refractive index of the Cerenkov generator - solid, liquid, or gas. The result of this effect is that particles below the energy threshold create no photon emission; particles above the threshold do generate a photon emission which becomes greater as the initial energy of the particle becomes further separated from the threshold energy. The threshold energy for electrons (betas) in any medium can be approximated using Equation 1:

$$E_T = 511 \left(\frac{1}{\sqrt{1 - \frac{1}{n^2}}} - 1 \right) \quad (1)$$

where E_T = threshold energy (KeV)

n = refractive index

The actual threshold energy cannot be calculated exactly using Equation 1 because, in real systems, the change of refractive index as a function of wavelength must be considered along with the spectral sensitivity of the experimental photon detector. However, one sees in Table I that the electron energy threshold varies over a rather narrow range (using n_D) for a considerable change in refractive index. Thus, for practical purposes, E_T at n_D is fairly realistic. But, it should be noted that some materials exhibit significant changes in refractive index in spectral regions (UV) where phototubes are sensitive. In these cases, major shifts can be observed in the energy threshold (6).

One of the best examples of an application that takes advantage of the energy threshold effect is a demonstration of the rapid determination of phosphorus-32/33 mixtures by Brown (7). Analysis of these mixtures by liquid scintillation counting does not offer a sufficiently high degree of sensitivity or accuracy because of the relatively low energy resolution inherent with this method. However, in an aqueous Cerenkov system, phosphorus-33 lies below the energy threshold and thus yields no response at all; phosphorus-32, which lies well above the threshold gives a

Table 1

CERENKOV ENERGY THRESHOLD FOR ELECTRONS IN VARIOUS SOLVENTS

Solvent	n_D	Threshold (KeV)
ethyl trifluoroacetate	1.31	283
water	1.33	263
ethyl alcohol	1.36	243
40% sucrose	1.40	219
toluene, 84% sucrose	1.50	174
bromoform	1.60	145
methylene iodide	1.76	111

strong reaction. Any counts observed in a phosphorus-32/33 mixture must be due only to the -32 contribution. (The response of -33 was experimentally determined to be <0.01%). By combining a Cerenkov count with a liquid scintillator count, Brown was able to determine mixtures of the phosphorus isotopes over an extremely wide range with errors usually less than 1.5%. The absolute accuracy of this method is limited mainly by the degree of uncertainty of phosphorus-33 liquid scintillation and the phosphorus-32 Cerenkov counting efficiency determinations.

The above example must be considered to be a "classic" since only one nuclide in the mixture gives any Cerenkov response while the other is completely masked. Nevertheless, many practical measurement systems have been developed where two (or more) nuclides have been assayed by taking advantage of the low detection efficiency of nuclides whose maximum beta energy lies above the threshold, but not significantly above. For example, thulium-170 with a beta energy maximum of almost 1 MeV counts with only about 5-6% efficiency. Using similar techniques, Randolph (8) has determined mixtures of Sr-89/90; Buchtela and Tschurlovits (9), Sr-89/90 and Y-90; and Ross (10), mixtures of Cl-36 and P-32. In each of these examples, some Cerenkov response was observed from every nuclide. The differences were sufficiently different, however, so that errors generally in the range of 5% were obtained over the span of mixture concentrations that were studied. The energy threshold technique is clearly not strictly limited to the theoretical cut-off value.

Chemical quenching

If any experimental parameter could be said to play a pivotal role in the success or failure of liquid scintillation counting, it must certainly be that of color and chemical quenching. The quenching phenomenon must be considered in virtually every aspect of the sample measurement - selection of cocktail formulation, sample preparation, instrument settings and correction procedures, etc. In almost one half of the liquid scintillation papers published in the last two years, a quenching problem and its solution was a major aspect of the investigation. In Cerenkov counting the situation is somewhat different; there is no known mechanism for chemical quenching to occur - it does not exist. This simple fact has virtually revolutionized the techniques used for sample preparation. Powerful chemical treatments that would be precluded from use with scintillation samples can be applied freely in Cerenkov systems. Dissolutions in strong acids, rapid molten salt fusions, and use of strong oxidants are all acceptable techniques. This new freedom for sample selection and processing has made a large impact on the recent growth of Cerenkov counting applications.

Although it is still true that color quenching effects can exist (and are in fact quite important), the ability to use strong chemical methods to treat samples can often completely eliminate color problems. A recent publication (11) describes the destruction of blood platelets by treatment with 0.5N sodium hydroxide and a subsequent direct count for P-32. The authors report that although the absolute counting efficiency is reduced from that obtained with scintillators, the advantages of a reduction in time, material, and expense are realized. They also note that the ease of sample recovery from the counting solution commends the approach to studies requiring subsequent analyses of the platelet lysate. Other recent examples of specific sample treatment include plant extracts and parts (12,13), gels (12), and biological materials in general (14).

Although forceful handling of samples to remove color quenching is possible, some investigators opt for a quench correction technique when the quenching is not severe. Virtually all of the usual correction methods have been examined and, contrary to a report by Elrick and Parker (15), the channels ratio may be among the best (16, 17, 18, 19). An alternate solution that has been proposed to circumvent color problems is the use of a "Cerenkov insert" as described by Ballance and Johnson (20, 21). Here, the sample is contained in a small, opaque container and dipped into the Cerenkov generating medium. Energetic particles pass through the container and into the solution that creates the Cerenkov emission. Since the colored sample is not dissolved or dispersed in solution, color effects are avoided. However, here it must be asked what this technique actually accomplishes. The sample, which is a self-contained entity, is no longer free of self absorption effects. The result is that reduced counting efficiencies are observed when compared to homogeneous counting. The small size of the insert also obviates against another benefit of Cerenkov methods, the use of large samples for lowered detection limits. It would appear that any advantage gained by use of the insert could also be realized by substituting a conventional liquid scintillator for the Cerenkov medium, with a considerable enhancement of the detection efficiency. Thus it is difficult to discern the application area of this method.

Spectral distribution, fluors, and counting efficiency

Since Cerenkov emission is solely the result of a rapid de-acceleration of a charged particle in a transparent medium, the use of conventional scintillation fluors is not necessary. This is the essential reason that chemical quenching does not occur in pure Cerenkov systems - there is simply nothing to chemically quench. The theoretical spectral output from a transparent medium ranges between its spectral transmission limits over the range where the refractive index is greater than one. The relative spectral yield varies as $1/\lambda^3 d\lambda$ and is the reason that the familiar Cerenkov glow in swimming pool reactors always appears blue to the eye. Wider range detectors show that even greater emission occurs in the ultraviolet spectral region as predicted by theory.

The large majority of phototubes used for scintillation counting exhibit a significant drop in sensitivity in the ultraviolet region and, thus, much of the energy from a

given particle event is not utilized for detection. For
high-energy particles, this is not a significant problem;
sufficient visible (and near UV) photons are generated to
reliably trigger the phototube and counting circuits.
However, for particles only marginally above the energy
threshold, there is not a sufficient number of detectable
photons and the observed counting efficiency is reduced
proportionately. It appeared obvious to many investigators
that the use of a wavelength shifter in the Cerenkov
medium could be used to recover and use much of the wasted
ultraviolet radiation. The use of such a material would
also destroy the directional characteristic of the emission
and further enhance the detection efficiency (22).

Early work by Herberg and Marshall (23) demonstrated
two materials that gave a significant gain in water. The
better of the two, 2-amino-6, 8-naphthalenedisulfonic acid,
gave an increase in pulse height of 30%. This compound
has been widely used since its utility was first
demonstrated. Another significant increase in pulse
height (as much as 100%) was shown to result from the use
of 4-methylumbelliferone by Porter (24). A large number
of other water soluble compounds have been proposed with
various levels of success; many of these materials were
examined in detail with regard to efficiency gain, stability,
and other experimental parameters (22).

An important comment must be made at this point.
Although the use of a wavelength shifting compound can
result in a significant increase in counting efficiency
in ideal systems, in practice, one must exercise a great
deal of care when they are used. The reason for this is
that once such materials are added to the Cerenkov
generator, the system becomes a hybrid Cerenkov-scintilla-
tion medium. As a result, many of the advantages described
earlier cannot be attained. The most important factor is
the realization that chemical quenching of the waveshifter
can occur. Thus, much of the freedom in sample pre-treatment is lost and, along with it, the convenience of the
"pure" Cerenkov method. Is there a way to "... have your
cake and eat it too?" The answer is yes.

MATERIALS AND METHODS

Baillon, et.al., (25) have demonstrated that an
increase by about a factor of 3.4 in photoelectron emission
from Cerenkov light can be obtained by coating a standard

Figure 1

Quartz counting vial with isolated wavelength shifter
compartment for Cerenkov counting

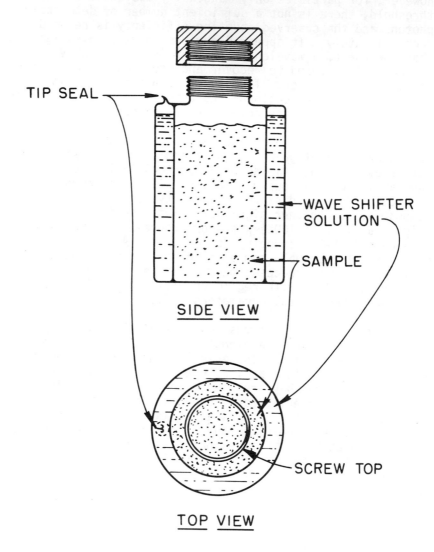

TIP SEAL

WAVE SHIFTER
SOLUTION

SAMPLE

SIDE VIEW

SCREW TOP

TOP VIEW

QUARTZ SAMPLE VIAL

photomultiplier tube with a thin deposit of a wavelength
shifter (lithium fluoride). Although the coating operation
is straightforward and requires only simple manipulation,
it does not appear to be a feasible approach for a general
purpose liquid scintillation counting system. This work,
however, suggested the design of a new type of counting
vial specifically for Cerenkov measurements using a waveshifter. The design of the vial is shown in Figure 1.
The device consists of two concentric chambers. The outer
chamber is used to contain a wavelength shifting solution
that may be either organic or inorganic. In this work,
the opening to the chamber was fitted with a stopper so
that various waveshifting solutions could be tested. The
inner chamber is used to hold the counting sample - about
10 ml maximum. The entire assembly is made of quartz so
that efficient transfer of the ultraviolet radiation from
the inner chamber to the shifter can take place. Since
the quartz vial is definately not a throw-away item,
the inner chamber of the vial is lined with a thin ($\sim$ 1.8
mil) polyethylene bag. The use of this liner prevents
radioactive contamination of the internal chamber surface
yet permits a high degree of UV transmission. The
construction of a similar all polyethylene vial appears
feasible but this fabrication was not attempted.

Scintillators and other materials used in this study
were scintillator or reagent grade. The strontium-89
radionuclide was prepared by irradiation of enriched
strontium-88 nitrate (98% enrichment) in the Oak Ridge
Research Reactor. The initial standardization of the tracer
was carried out by $4\pi\beta$ counting techniques. A Packard
series 3000 liquid scintillation counter was used for the
Cerenkov measurements. Maximum amplifier gain and wide
window settings were used.

RESULTS

The strontium-89 nitrate tracer was dissolved in 0.1 $\underline{N}$
HCl. An aliquot of this solution was added to 10 ml of
0.01 $\underline{N}$ HCl and this sample was placed into the inner chamber
of the vial and sealed with a screw cap. The aliquot was
selected so that about 30,000 dpm were contained in the
sample. The outer chamber was filled with various solutions
with and without an added waveshifter. A count of the sample
was made so that a total of at least 50,000 counts were
collected in each test. The relative counting efficiency

was determined for each counting combination; the results are summerized in Table II.

Table II

RELATIVE COUNTING EFFICIENCY OF Sr-89
USING AN ISOLATED WAVELENGTH SHIFTING VIAL

Waveshifter Chamber Contents	Concentration	Relative Counting Eff.
water	100%	1.00
ethanol	95%	1.02
benzene	100%	1.06
toluene	100%	1.07
dimethyl POPOP	0.7 gm/l in toluene	1.88
4-methyl- umbelliferone	0.5 gm/l in 75% ethanol	1.69
2-naphthol-3, 6-disulfonic acid-sodium salt	0.1 gm/l in water	1.59
β-naphthol	0.1 gm/l in 50% ethanol	1.37
1-naphthylamine	0.1 gm/l in 50% ethanol	1.22

Three strong reagents (and powerful chemical quenchers in conventional scintillation mixtures) were added to the sample compartment during the tests with umbelliferone and dimethyl POPOP. One milliliter of concentrated sulfuric acid, 10% sodium hydroxide, and 30% hydrogen peroxide were each tested in separate experiments. In every case, no significant change in the relative counting efficiency was observed.

Another series of tests was designed to examine the effect of color quenching in the sample. Here, either the wavelength shifter dimethyl POPOP in toluene or plain water was used in the external compartment of the Cerenkov vial. Dye solutions were added to the sample compartment until a strong color component was visible. However, no attempt was made to determine the actual absorbances of the solutions since relative measurements were being made. Coloring agents used were water-soluble FD&C coal-tar dyes. The results obtained in these tests are shown in Table III.

Table III

THE EFFECT OF COLOR IN THE SAMPLE
WHEN USING THE CERENKOV VIAL

Color & λ abs.$_{max}$	Waveshifter	Relative Counting Eff.
None	None (Water)	1.00
FD&C yellow (400 nm)	None	0.62
FD&C yellow	DMPOPOP	0.91
FD&C red (510 nm)	None	0.57
FD&C red	DMPOPOP	1.09
FD&C blue (625 nm)	None	0.91
FD&C blue	DMPOPOP	1.83

all tests with Sr-89 tracer

DISCUSSION

The results presented in Table II clearly demonstrate that the specially constructed Cerenkov vial can be used to significantly increase the detection efficiency of beta emitters by separating the sample and the wavelength shifting component. In the case of Sr-89, which is a moderately energetic emitter (E_{max}=1.49 MeV), a major increase is observed. Since previous work has indicated that waveshifting becomes more crucial with weaker emitters, an even greater increase could be projected for nuclides such as Cl-36 or Tm-170. A vital consideration is that use of the new vial design does not change the detection process from a true Cerenkov function. Thus, insensitivity to chemical quenching is preserved and virtually any form of sample dissolution and pre-treatment can be employed. Another prominent feature of the new vial is that the wavelength shifter can be used in a totally organic system.

This results in the ability to select more efficient fluor
materials and those that match phototube spectral response
more closely.

An additional aspect of the vial concept is the
improved response when some visible color quenching remains
in the sample (Table III). This is important because some
biological materials resist complete decoloration using
simple procedures. It is only fair to note that ultraviolet
absorbers in the sample will clearly reduce the effectiveness of this technique. But, we have found that many of
these materials are easily destroyed during the routine
initial sample dissolution. The only real disadvantage
of this method is that the volume of sample that can be
conveniently handled is reduced by about a factor of two.
However, in the majority of situations, sample size is
not a limiting factor.

REFERENCES

1. E. H. Belcher, Proc. Roy. Soc. A216, 90 (1953).

2. J. V. Jelley, Cerenkov Radiation, New York: Pergamon
    Press (1958).

3. V. K. Haberer, Atomwirtschaft 10, 36 (1965).

4. B. Francois, Int. J. Nuc. Med. and Biology 1, 1 (1973).

5. H. H. Ross, Anal. Chem. 41, 1260 (1969).

6. H. H. Ross in The Current Status of Liquid Scintillation
    Counting, p. 123 (E. D. Bransome, Ed.). New York and
    London: Grune and Stratton (1970).

7. L. C. Brown, Anal. Chem. 43, 1326 (1971).

8. R. B. Randolph, Int. J. Appl. Radiation and Isotopes
    26, 9 (1975).

9. K. Buchtela and M. Tschurlovits, ibid. 26, 333 (1975).

10. H. H. Ross and G. T. Rasmussen in Liquid Scintillation
    Counting, p. 363 (P. E. Stanley and B. A. Scoggins, Eds.).
    New York and London: Academic Press (1974).

11. R. D. Smith, J. J. B. Anderson and M. Ristic, Int. J. Appl. Radiation and Isotopes 23, 513 (1972).

12. F. Fric and V. Palovcikova, ibid. 26, 305 (1975).

13. S. Seshadri, ibid. 26, 557 (1975).

14. A. Lauchli in Organic Scintillators and Liquid Scintillation Counting, p. 771 (D. L. Horrocks and C-T. Peng, Eds.). New York and London: Academic Press (1971).

15. R. H. Elrick and R. P. Parker, Int. J. Appl. Radiation and Isotopes 19, 263 (1968).

16. A. T. B. Moir, ibid. 22, 213 (1971).

17. L. I. Wiebe, A. A. Noujaim and C. Edis, ibid. 22, 463 (1971).

18. J. E. Johnson and J. M. Hartsuck, Health Phys. 16, 755 (1969).

19. R. D. Stubbs and A. Jackson, Int. J. Appl. Radiation and Isotopes 18, 857 (1967).

20. P. E. Ballance and S. Johnson, Planta 91, 364 (1970).

21. P. E. Ballance and S. Johnson, Health Phys. 20, 447 (1971).

22. H. H. Ross in Organic Scintillators and Liquid Scintillation Counting, p. 757 (D. L. Horrocks and C-T. Peng, Eds.). New York and London: Academic Press (1971).

23. E. Heiberg and J. Marshall, Rev. Sci. Instrum. 27, 618 (1956).

24. N. Porter, Nuovo Cim. 5, Series 10, 526 (1957).

25. P. Baillon, et al., Nuc. Instrum. Methods 126, 13 (1975).

METHYL SALICYLATE AS A MEDIUM FOR RADIOASSAY OF
[36]CHLORINE USING A LIQUID SCINTILLATION SPECTROMETER

L.I. Wiebe and C. Ediss
Division of Bionucleonics and Radiopharmacy
University of Alberta
Edmonton, Canada

Abstract

Methyl salicylate (MS), a high refractive index liquid
with wave-shifting properties, has been used as a Cerenkov
radiation generating medium for the radioassay of ^{36}Cl by
liquid scintillation (LS) spectrometer. Comparative
experiments, using both a standard toluene-based LS fluor
and toluene alone, for the measurement of ^{36}Cl were undertaken. The methyl salicylate medium was found to perform at
an intermediate counting efficiency, near that for the LS
fluor. In the presence of moderate amounts of nitromethane,
the MS was less susceptible to chemical quenching effects
than either the fluor or toluene.

Counting efficiencies for ^{36}Cl in MS, toluene and
toluene fluor respectively were 82.4, 28.4 and 100.3 percent
with a Picker Liquimat 220 LS spectrometer, and 91.6, 54.9
and 100.0 percent with a Searle Mark III LS spectrometer.
The addition of nitromethane (11.3 percent of final volume)
reduced these efficiencies to 50.5, 10.0 and 15.4 percent,
and to 58.8, 12.4 and 19.0 percent, respectively.

The data are discussed in relation to observed changes
in the pulse height spectra. Chemical quench correction by
ESCR and by SCR methods is reported.

Introduction

The use of liquid scintillation (LS) counters for the
measurement of Cerenkov radiation emitted upon the decelera-
tion of highly energetic β particles in liquid media has
become increasingly popular over the past several years.[1]
The technique has found particular acceptance as an alternate to LS counting of samples which contain appreciable
quantities of substances which chemically quench the LS

93

process. Many liquids have been used as Cerenkov generating
media, particularly water, and organic solvents with refractive indices considerably greater than the refractive index
of water. In addition, Cerenkov generating systems containing wave-shifting chemicals have been developed. These
systems, although no longer purely Cerenkov generators, have
provided improved counting efficiencies.[1,2]

This paper describes the performance characteristics of
methyl salicylate (MS-C) as a Cerenkov generator for the
radioassay of [36]Chlorine. This high refractive index liquid
posesses wave-shifting properties, absorbing the ultraviolet
emissions of the Cerenkov spectrum, and re-emitting photons
in that visible portion of the electromagnetic spectrum to
which the photomultiplier tubes are more sensitive. Comparisons are drawn to the characteristics of Cerenkov counting
of [36]Cl in toluene (T-C), and to LS counting of [36]Cl in
toluene containing PPO and POPOP.

Experimental

Methyl salicylate (laboratory grade, Fisher Scientific
Co., Edmonton) was distilled before use. Scintillation
grade toluene, PPO and POPOP (Fisher Scientific Co., Edmonton) were used as supplied. The LS fluor was prepared to
contain PPO (4 g l^{-1}) and POPOP (50 mg l^{-1}) in toluene.

Commercial LS glass vials (Kimble, Toledo, Ohio) were
used for all procedures. Ten ml of the appropriate counting
medium was pipetted into each vial; these vials were then
scintillation counted to assure absence of contamination and
photo or chemiluminescence effects. Standard (±3%) [36]Cl-
Chlorobenzene (50 µl; 4.17 x 10^5 dpm ml^{-1}; New England
Nuclear, Boston) was added to each vial by microsyringe,
after which the individual vials were scintillation counted
again to determine the absolute counting efficiency. The
LS counting of each vial was repeated after each subsequent
addition of small volumes of nitromethane (Reagent grade,
Fisher Scientific Co., Edmonton).

Vials were scintillation counted in a Liquimat 220 LS
spectrometer (Picker Nuclear, New York) and in a Mark III
LS spectrometer (Searle Instrumentation, Chicago). Pulse
height spectra obtained from the Liquimat 220 LS spectrometer were stored in either a Nuclear Chicago Model 25601
multichannel analyzer (MCA) or in a Northern Scientific
NS 636 MCA.

Results
      Methyl salicylate was found to give high counting
efficiency for ^{36}Chlorine in the absence of chemical
quenching agents using both the 'new generation' Mark III
LS counter, and the older Liquimat 220 instrument. The
superiority (E^2/B) of MS over the standard LS fluor, although
apparent in the absence of nitromethane quench, became
increasingly evident as the concentration of nitromethane
exceeded 1 percent. At maximum quench, MS-C counting was
three times as efficient as LS counting. The T-C counting
was much less efficient than either the LS or MS-C process,
with or without chemical quench. Of particular interest
was the large decrease in counting efficiency upon addition
of small amounts (10 μl; 0.1%) of nitromethane. Counting
efficiencies, background count rates, and E^2/B values, are
shown for the LS, MS-C and T-C systems using both liquid
scintillation counters, in Table 1.
      Quench correction by 'external standard pulse' (ESP)
and sample channels ratio (SCR) was investigated using the
programmed 'windows' of the Mark III spectrometer. The
ESP method was found to perform well in moderately quenched
LS samples, but to be unsatisfactory for MS-C and T-C
samples. The dynamic range of SCR provided by programmed
windows was found to be adequate for the entire range of
quench in MS-C and T-C samples, but was very limited for
LS samples in both programs 2 (^{14}C) and 10 (^{36}Cl). In
addition, the use of program 10 led to the exclusion of
large numbers of pulses, resulting in unnecessarily low
counting efficiencies at severe quench levels. Quench
correction data are given in Figures 1 and 2.
      The pulse height spectra (PHS) depicted in Figure 3
show major differences from each other. The comparatively
broad PHS of the MS-C system is noteworthy. The ^{137}Cs
compton electron induced Cerenkov spectrum in MS (Figure 4)
is seen to match closely the unquenched MS-C ^{36}Cl PHS. The
external standard count rate (∿ 120,000 cpm) observed with
the unquenched MS-C generator was adequate to provide
acceptable statistical precision. Even with the significant
loss of counts with severe quenching the method would be
useful. External standard channel ratios obtained are
listed in Table 2.

Discussion and Conclusions
      The efficacy of methyl salicylate as a liquid Cerenkov
radiation generating medium for ^{36}Cl has been shown to
approach that of a typical LS system in terms of unquenched
sample counting efficiency, and to surpass the LS system

Table 1. Observed counting efficiencies for ^{36}Cl in LS fluor (PPO/POPOP/toluene), methyl salicylate (MS-C) and toluene (T-C) with increasing quantities of nitromethane. Measurements were made with the Picker Liquimat 220 and the Searle Mark III LS spectrometer.

Volume CH$_3$NO$_2$ (µl)	LS		MS-C		TC	
	220[1]	Mk III[a]	220[2]	Mk III[b]	220[3]	Mk III[c]
0	100.3	100.0	82.4	91.6	28.4	54.9
10	100.0	99.6	81.6	90.5	19.4	30.7
20	100.0	99.4	80.9	89.7	17.3	25.3
40	98.4	98.9	79.4	88.3	16.0	21.6
80	94.5	96.4	77.2	86.3	14.3	19.0
160	85.4	86.8	74.3	83.0	13.4	17.1
320	54.5	61.4	69.5	79.6	12.7	15.6
640	27.2	32.0	61.6	70.0	11.6	14.2
1280	15.4	19.0	50.5	58.8	10.0	12.4
B	141.0	48.8	88.7	25.0	46.0	24.4
E^2/B	71.3	204.9 572.6[d]	76.5	335.6	17.5	123.5

1. window 10-1000
2. window 10-650
3. window 10-450

a. P-32 window, Program 4
b. C-14 window, Program 2
c. H-3 window, Program 1
d. Cl-36 window, Program 10

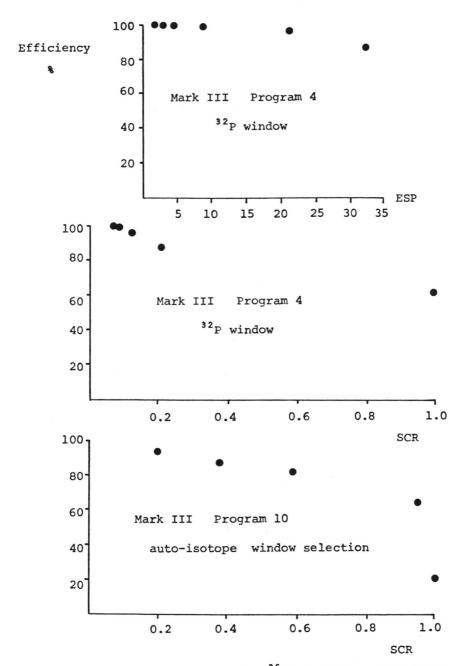

Figure 1. Quench correction for ^{36}Cl in Toluene/PPO/POPOP using the ESP and SCR methods.

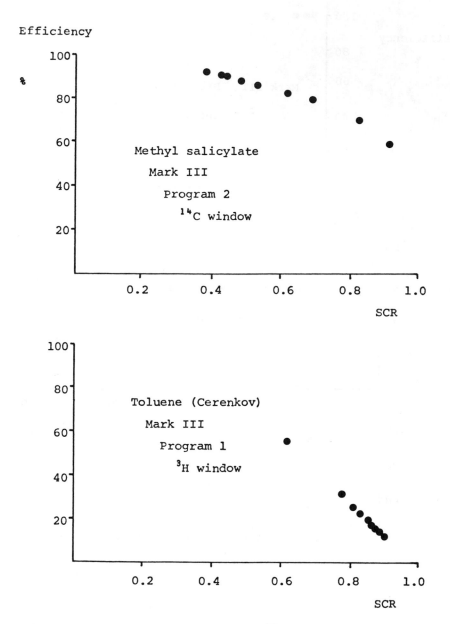

Figure 2.   Quench correction for ^{36}Cl in methyl salicylate (MS) and toluene (T) using the SCR method.

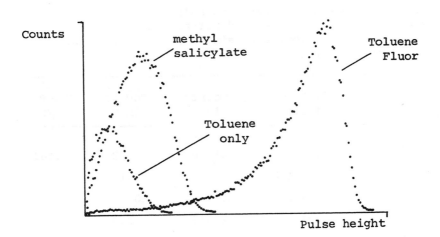

Figure 3. Pulse height spectrum of ^{36}Cl in various media.

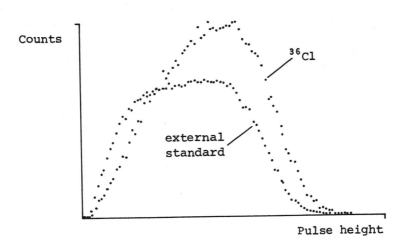

Figure 4. Pulse height spectrum in methyl salicylate of ^{36}Cl and the Liquimat 220 ^{137}Cs external standard.

Table 2. External standard channels ratio (ESCR), external standard counts, and percent efficiency for ^{36}Cl in methyl salicylate, using the Picker Liquimat 220 LS spectrometer.

Volume of CH_3NO_2	Relative % efficiency	Counts in D	Counts in C	ESCR D/C
0	100.	31558	25772	1.241
10	97.4	29234	26417	1.105
20	96.1	29228	28086	1.058
40	94.3	26546	28320	0.937
80	90.8	24564	29152	0.842
160	85.3	20526	29303	0.742
320	78.1	16802	29259	0.607
640	68.3	11439	27065	0.431
1280	53.1	6648	23386	0.287

D = 125-420

C = 10-180

with respect to resistance to severe chemical quench.
Examination of the PHS for ^{36}Cl in MS and toluene suggests
that the high counting efficiencies observed in MS are
attributable to the wave-shifting characteristics of that
liquid. The photon wave-lengths are shifted from the ultraviolet region to the visible region, thereby decreasing
absorption of photons by the solvent and by the glass of
vials and photomultiplier tube (PMT) faces. Furthermore,
wave-shifted photons have isotropic spacial distribution,
thereby increasing the detection efficiency in typical LS
spectrometers in which the PMTs are at an angle of 180° to
each other and are operated in coincidence.[3] The higher
refractive index of MS (1.522 vs 1.494 for toluene)[4] should
also increase the photon yield[5] by decreasing the Cerenkov
threshold level. Inspection of the ^{36}Cl beta energy spectrum reveals that approximately 65% of the β^- particles have
energies above the calculated Cerenkov threshold for β
particles in MS. In view of the very high counting efficiency that has been observed (91.6%) it would appear that
the refractive index of MS at the Cerenkov wavelengths is
indeed greater than 1.522 (R.I. Na_D^{20}).

Quench correction by SCR has been found to be satisfactory in the MS-C and the T-C systems using the programmed
windows (2-^{14}C; 1-^{3}H, respectively) of the Searle Mark III
LS spectrometer. The small dynamic range observed for both
ESCR and SCR methods in the LS analysis could undoubtedly
be remedied by more appropriate window selection than that
presented. The loss of counts experienced when using a ^{36}Cl
LS standard to set Program 10 windows of that spectrometer
could similarly be reduced. The use of the SCR method with
the Liquimat 220, although not investigated, would in all
likelihood be useful for quench correction through appropriate channel selection. Quench correction by ESCR using
^{137}Cs external standard should be ideal because of similarities in the PHS for ^{137}Cs compton electrons and ^{36}Cl β
particles in MS.

As pointed out previously[1], the inclusion of wave-
shifting compounds in Cerenkov generating media produces a
system which is no longer purely Cerenkov in nature. In
MS-C counting, however, it is not apparent that significant
chemical quenching does occur, as the rate of decline of
counting efficiency remains at 1% or less per 10 µl of
nitromethane added, up to the maximum added (1280 µl per
10 ml MS). This small rate of decline may in fact be
primarily due to a decrease in the refractive index of MS
(RI = 1.522) by addition of nitromethane (RI = 1.380[4]), and
the absorption of photons by nitromethane.

The T-C data show a similar slow decline in counting efficiency, but only after addition of 10-20 µl of nitromethane. In this case, it would appear that the initial large decreases in counting efficiency are due to chemical quenching of the scintillation process in toluene, a solvent which has an appreciable photon yield itself.[6] Subsequent decreases (after 20 µl of nitromethane) are in the order of 1% or less, similar to those observed in MS.

References

1. H.H. Ross and G.T. Rasmussen, in Liquid Scintillation Counting: Recent Developments, P.E. Stanley and B.A. Scoggins (Eds.), Academic Press, N.Y. (1974) p. 363.
2. R.P. Parker and R.H. Elrick, in The Current Status of Liquid Scintillation Counting, E. Bransome (Ed.), Grune and Stratton, N.Y. (1970) p. 110.
3. H.H. Ross, in Organic Scintillators and Liquid Scintillation Counting, D.L. Horrocks and C.T. Peng (Eds.), Academic Press, N.Y. (1971) p. 757.
4. Handbook of Chemistry and Physics, R.C. Weast and S.M. Selby (Eds.), The Chemical Rubber Co., Cleveland, Ohio, 47th Edition (1966/67) p. E151.
5. I.M. Frank and I.G. Tamm, Dokl. Akad. Nauk. SSSR 14, 109 (1937).
6. D.L. Horrocks, Applications of Liquid Scintillation Counting, Academic Press, N.Y. (1974) p. 19.

SOME FACTORS INFLUENCING

EXTERNAL STANDARDIZATION

Philip E. Stanley,

Department of Clinical Pharmacology
The Queen Elizabeth Hospital
Woodville, South Australia 5011

ABSTRACT

A study has been made of some factors which effect the external standard ratio-efficiency curve for samples counted in a Searle Isocap 300 liquid scintillation spectrometer fitted with lesser pulse height analysis. The effect of changing the volume of a simple scintillation solution was investigated and it was shown that little difference occurs in the external standard ratio-efficiency curve when chemically and colour quenched tritium and chemically quenched ^{14}C-samples were studied whereas substantial differences were apparent for colour quenched ^{14}C-samples. In addition a study has shown that the percentage standard deviation of the external standard ratio for background samples under a range of conditions is less than 0.75% while for samples containing high activity (10^6 DPM) the value rises as high as 2.8%.

INTRODUCTION

Although automatic external standardization was introduced more than a decade ago (1-2) there is still considerable reluctance among some users of liquid scintillation spectrometers to employ this useful technique to assess the counting efficiency of their samples and there is yet another group who use it in less than the

proper manner. Some of the reticence is no doubt well
founded firstly because of certain shortcomings in the
mechanisms of early units and a lack of understanding of
the technique which led to experiences which are not readily
dispelled even though modern units are very reliable.
Secondly, some publications describe the use of the
procedure for samples which are heterogeneous or slowly
soluble or which adhere or are adsorbed to the glass wall
of the vial. Under such conditions the method can give
very unreliable results. In addition, the correct use of
external standardization with scintillation solutions
containing surfactants requires that the standard curve be
generated using a quenching agent which is the same or very
similar to that encountered in the sample. In addition the
proportions of sample, surfactant and scintillation solution
should be carefully controlled if reliable results are to
be obtained. Thirdly, possibly as a result of commercial
promotion, the procedure of external standardization is
often considered to be "independent" of sample volume and
quenching agent. Thus workers sometimes do not ascertain
that this is indeed the case for their samples and the
spectrometer they use. Fourthly, some workers use one
calibration curve, derived perhaps from the set of quenched
standards supplied with a new instrument, and apply it to
all samples for many months (or years) without revalidating
it or checking for instrument drift. The particular
scintillation solution used in the standards and its
resistance to quenching may be quite different from the one
the worker is currently using. The results which are
generated are thus likely to be incorrect. Fifthly, the
nature of the processes which give rise to chemical and
colour quenching are quite different (3-7) and the external
standard ratio vs counting efficience must be investigated
for the instrument being used, the settings of the pulse
height analyzers and the quenching agents. Sometimes two
quite different curves arise especially with isotopes with
β-energies similar to or greater than carbon-14.

Since there is a dearth of published information on this
subject the present investigation was undertaken to look
carefully at a few of these areas, namely, the effect of
the volume of the sample on its counting efficiency and
external standard ratio for both the chemical and colour
quenched samples containing tritium or carbon-14. In
addition a study was made of the reproducibility of the
external standard ratio for samples having no activity as

compared with those containing very high activity and assessment was made of the reproducibility of the external standard ratio value and its relationship to the geometry of the scintillation vial.

MATERIALS

Instrument

The liquid scintillation spectrometer used was a Model 6872 Searle Analytic Isocap 300 with a temperature controlled counting cabinet operated at 20.5^0C. The instrument was fitted with an external standard consisting of a barium-133 source positioned automatically beneath the sample vial using a rigid tungsten rod. Live timing and synchronous counting facilities were fitted to the spectrometer. The latter causes the timer to continue, after a preset count has been reached, until the next increment of the printed time has elapsed. This permits more accurate assessment of counting rates for high activity samples especially when they are measured for short preset times.

The liquid scintillation spectrometer was fitted with a photon monitor, cross-talk discriminators (8) and pulse analysis was carried out by the lesser pulse height procedure (9,10). The two analyzers were set as follows:-
TRITIUM - CAPS 4 Ch. A = 0.5 $\rightarrow$ 9 and Ch. B = 0.5 $\rightarrow$ 18 keV, external standard ratio channels (ESR) Ch. A = 5.6 $\rightarrow$ INF and Ch. B = 0.5 $\rightarrow$ 18 keV.
CARBON-14 - CAPS 5 Ch. A = 1.2 $\rightarrow$ 60 and Ch. B = 4.5 $\rightarrow$ 150 keV, ESR Ch. A = 0.5 $\rightarrow$ 18 keV and Ch. B = 3.6 $\rightarrow$ INF keV.
The instrument was calibrated (CAPS L position) two days before this study commenced and the calibration was checked daily over the four weeks during which the work was performed. The instrument was used exclusively for this study and no change in calibration was required.

Scintillation solution

The scintillation solution contained 8.00 g PPO made up to one litre with toluene. Its density at 20^0C was taken as 0.866 g/ml. PPO (M.Pt. = 71^0C) was obtained from Ajax Chemicals, Australia, and toluene (Boiling Range 109.5 - 111.5^0C) was Pronalys Grade and obtained from May and Baker, Australia. Both were used without further purification.

Quenching Agents

Nitromethane, was obtained from Ajax Chemicals Australia
and used as the chemical quenching agent. Its density was
taken to be 1.13 g/ml.

The colour quenching agent was β-carotene, obtained from
Koch-Light Ltd. U.K., prepared as a 20 mM solution in the
standard scintillation solution. Its density was measured
crudely to be 0.868 g/ml. Prior to use it was filtered
through a Whatman No. 1 paper. β-carotene was chosen in
preference to the more commonly used methyl orange since the
latter gives solutions which are not optically stable.

Standards

Tritiated hexadecane and ^{14}C-hexadecane were obtained
from The Radiochemical Centre, Amersham, U.K. and has
activities with estimated overall uncertainties of ± 3% and
± 2% respectively. The density of hexadecane was taken as
0.773 g/ml and 20^{0}C. For the short study involving high
activity samples, large volumes (up to 1.5 ml) of hexadecane
could not be used since considerable dilution of PPO would
have occurred. Instead ^{3}H-toluene and ^{14}C-toluene standards
(± 1% and ± 3% respectively), obtained from Packard
Instrument Inc., U.S.A., were used and PPO was added to
obtain a concentration of 8 mg/ml (8 g/litre).

Scintillation vials

Glass scintillation vials were obtained from Packard
Instrument Inc. and these were fitted with foil lined lids.
Sufficient vials for the study were selected as being
similar on the basis of their outer dimensions and weight
(see reference 5 for details). Light transmission within
the batch had a standard deviation (S.D.) = 0.75% as judged
by counting rates observed when they contained 20.0 ml of a
standard scintillation solution spiked with tritium (10^{5}DPM)
and then counted to 800,000 counts. External standard
ratios using the tritium CAPS setting had a S.D. = 0.9%.
The vials were then washed before use in the present study.

METHODS

    Two volumes of scintillation solution were labelled, one with tritiated hexadecane and the other with ^{14}C-hexadecane. Standard hexadecane, placed in a small tared vessel, was weighed to within a 0.1 mg and then quickly transferred to 350 ml (checked by weighing) scintillation solution. Thus a solution containing an accurately known activity (about 10^5 DPM per 20 ml) was obtained. From each of these solutions ten aliquots of 30 ml were transferred to tared 50 ml glass tubes having close fitting stoppers. Speed was essential to avoid evaporation of the toluene. The tubes were then reweighed and any contents exceeding an error range of ± 0.15 ml were readjusted to come within this tolerance. Each set was used in the study of chemical quenching and after this had been completed (about two weeks) another ten aliquots of each standard scintillation solution were prepared and used in the investigation concerning colour quenching.

    Nitromethane and β-carotene solution were added as predetermined volumes to the 30 ml aliquots of scintillation solution. These volumes were derived from a pilot study and for nitromethane up to 150 µℓ for ^{3}H and up to 390 µℓ for ^{14}C were required while for the β-carotene solution the respective volumes were 72 µℓ and 155 µℓ. Gilmont precision microlitre pipettes were used for dispensing the quenching agents. Since these pipettes were calibrated "to contain" they were rinsed twice in the radioactive solution and then cleaned before being used for the next sample.

    A Kontes dispenser was used to transfer by glass pipette, 5 ml of the quenched standards to the selected and weighed (± 1 mg) scintillation vials which were then capped and reweighed to obtain the weight of the scintillation solution and hence the DPM of the hexadecane contained therein. Appropriate allowance in this calculation was made for the volume of added quenching agent. The vials were then temperature equilibrated in the spectrometer with its lid closed so as to preclude problems of chemi-luminescence or phosphorescence of the glass vials. Each sample was then counted at least twice for twenty minutes or 200,000 counts (1 σ = 0.22%) together with the external standard cycle appropriate to the CAPS setting. The vials were then reweighed and generally less than 10 mg loss was observed and it was therefore assumed that since

hexadecane has a very low vapour pressure the loss was
mainly due to toluene and perhaps a little nitromethane.
Consequently, it was taken that there had been no loss of
radioactivity or change in quenching. Using this new
starting weight a further 5 ml aliquot of the same quenched
standard was added so that the vial now contained 10 ml.
The counting process was then repeated and the overall
procedure reproduced with the same vials next containing
15 ml and then 20 ml of the quenched standard solutions.

High activity samples

A predetermined volume of scintillation solution was
placed in a weighed vial and then sufficient (0.5 → 1.5 ml)
of standard toluene (labelled either with ^{3}H or ^{14}C and
containing 8 mg/ml PPO) was added to give around 10^6 DPM.
Actual DPM was obtained by weighing. An air quenched
sample was evaluated (see next section) as were samples
which had been quenched with the nitromethane or β-carotene.
The total volume of sample was 20.0 ml.

Reproducibility of external standard ratio values

The scintillation spectrometer was first modified so
that when a sample was kept loaded it was possible to
obtain repeated evaluations of the external standard ratio.
A sealed unquenched standard background sample was used and
was kept in the detector in a constant position while thirty
external standard ratio evaluations were made using the
CAPS 4 (tritium setting). Then with the same CAPS setting
thirty external standard evaluations were performed with
the vial being loaded on each of thirty cycles of the
sample belt. Thus, in the first procedure the reproducibility of the ^{133}Ba external standard source could be assessed
since the geometry of the vial was kept constant while in
the second the variation the vial geometry, together with
the positioning of the external standard could be measured.
This procedure was repeated on the same samples using the
CAPS 5 setting for carbon-14.

A similar study was made for background samples which
had been either heavily chemical or colour quenched.

The high activity samples, prepared as described above,
were also subjected to a study in which the reproducibility

of the external standard ratio was assessed while the vial
was kept in the loaded position. Five observations were
made on each sample.

RESULTS AND DISCUSSION

The effect on the ESR-efficiency curve when different
volumes of sample were appraised is presented for tritium
in Fig. 1 (chemical quench) and Fig. 2 (colour quench) and
for carbon-14 in Fig. 3 (chemical quench and Fig. 4
(colour quench).

In the spectrometer used for this study the external
standard windows are very similar for tritium and carbon-14
(see *Instruments*) and the form of the ratio taken for
tritium is simply the reciprocal of that for carbon-14.
Thus the curves of the two radionuclides are related in an
inverse manner.

Consider first the results for tritium (Figs. 1 and 2).
All the points lie more or less along the same line with a
maximum spread of about ± 1.5% in the counting efficiency
for both the chemical and colour quenched samples. The
outlying points are mainly from the 5 ml sample size.
Thus for most purposes the curve for the chemically quenched
samples and for the colour quenched samples can be regarded
as being reasonably independent of sample volume between
5 and 20 ml. The two curves themselves are very similar
for sample efficiencies above 30% but below this value they
diverge and samples containing both chemical and colour
quenching agents would be difficult to assess with accuracy.

The results for the chemically quenched samples of
carbon-14 (Fig. 3) show that the ESR-efficiency curves for
the four different volumes can be completely super-imposed
and thus they can be considered as being independent of
sample volume between 5 and 20 ml. Like the tritium
samples the ESR values and efficiencies do change for the
various volumes of sample but the points simply move along
the curve.

Colour quenched samples containing carbon-14 (Fig. 4)
show a separate relationship between ESR and efficiency for
each of the four volumes studied. The curves are similar
for efficiencies above 70% but diverge significantly below

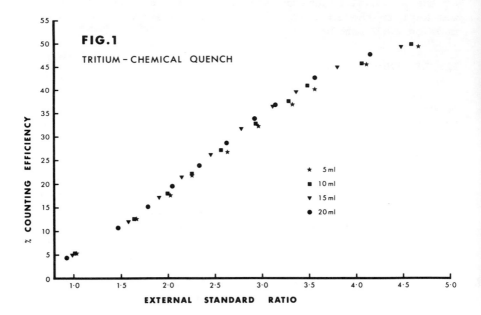

FIG.1

TRITIUM — CHEMICAL QUENCH

★ 5 ml
■ 10 ml
▼ 15 ml
● 20 ml

% COUNTING EFFICIENCY

EXTERNAL STANDARD RATIO

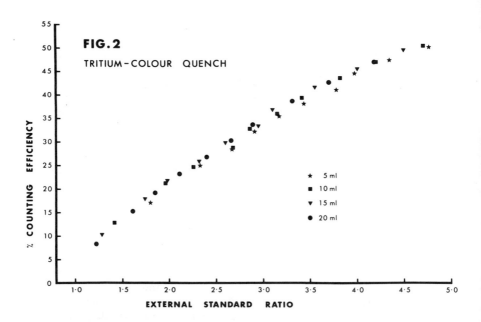

FIG.2

TRITIUM — COLOUR QUENCH

★ 5 ml
■ 10 ml
▼ 15 ml
● 20 ml

% COUNTING EFFICIENCY

EXTERNAL STANDARD RATIO

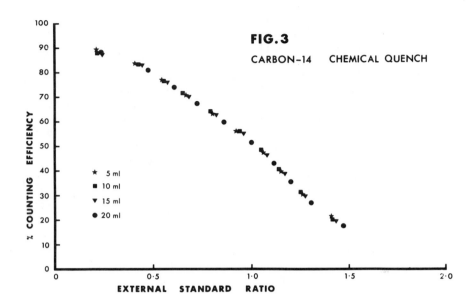

FIG. 3

CARBON-14   CHEMICAL QUENCH

★ 5 ml
■ 10 ml
▼ 15 ml
● 20 ml

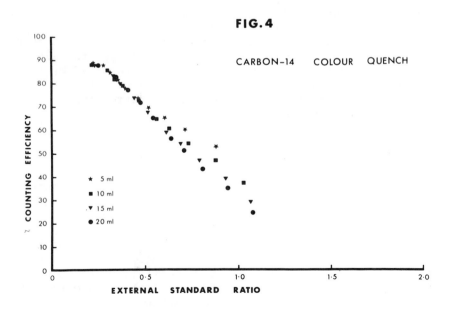

FIG. 4

CARBON-14   COLOUR   QUENCH

★ 5 ml
■ 10 ml
▼ 15 ml
● 20 ml

this value. Note that the ESR value for the tenth (most quenched) sample of 5 ml volume lies between the eighth and ninth sample of 20 ml volume. This extreme value would be read as an efficiency of 39% on the curve for 20 ml samples whereas in reality it is 53%.

Why then do the tritium samples not behave in the same manner as the carbon-14 ? The explanation hinges on two facts. Firstly for samples of different volumes the photons must travel different pathlengths to escape from the vial. In colour quenching the photons are absorbed by the agent and the absorbance is related exponentially to the path length. Secondly the energy of the radionuclide has a considerable influence since for an average energy (50.2 keV) carbon-14 β-event some 800 photons are produced whereas for tritium (average energy 5.6 keV around 65 are generated. The results generated from a computer simulation of the liquid scintillation process (5) indicate that for unquenched carbon-14 the number of photoelectrons produced at the photocathode of the photomultiplier is around 60 whereas for tritium the number is about 4. Thus for tritium, from the statistical viewpoint, little variation is possible in the pulse height since at least two coincident photoelectrons are required to enable coincidence and to register a single count. However for carbon-14 with 60 photons per average event there is more statistical variation within the pulse heights due to the stochastic nature of the overall liquid scintillation process and more particularly in this case with the path lengths travelled by the photons and whether or not they are absorbed by the colour quenching agent. Due to the larger number of photoelectrons there is also of course a better chance of enabling coincidence and registering a count.

Thus all colour quenched, carbon-14 samples should have the same volume as the standards. Since the chemical quench ESR-efficiency curve does not lie close to any of the colour quench curves the determination of efficiencies would appear to be precluded for ^{14}C-labelled samples containing both chemical and colour quenching agents.

The results of the study on the reproducibility of ESR on background samples is presented in Table I. It can be seen that the ratios obtained when the vials were kept in the loaded position are almost identical to those obtained when the vials were loaded once on each cycle of the sample

TABLE I

REPRODUCIBILITY OF ESR ON BACKGROUND SAMPLES

CAPS 4   (TRITIUM + ESR)

	Vial Kept Loaded		Vial Cycled	
	ESR	% S.D.	ESR	% S.D.
Unquenched	5.24	0.62	5.28	0.71
Heavy chemical quench	0.93	0.38	0.93	0.34
Heavy colour quench	1.28	0.42	1.28	0.48

CAPS 5   (CARBON-14 + ESR)

	Vial Kept Loaded		Vial Cycled	
	ESR	% S.D.	ESR	% S.D.
Unquenched	0.191	0.70	0.191	0.69
Heavy chemical quench	1.520	0.43	1.520	0.27
Heavy colour quench	1.020	0.26	1.020	0.31

ESR = External Standard Ratio
n   = 30

113

TABLE   II

REPRODUCIBILITY OF ESR ON SAMPLES CONTAINING 10^6 DPM

CAPS 4   (TRITIUM + ESR)

		% Efficiency	ESR	% S.D.
Air Quenched		52.0	4.42	2.84
Chemical Quenched	1	41.7	3.27	1.97
	2	32.8	2.69	0.95
	3	17.7	1.84	0.60
	4	5.0	0.94	0.50
Colour Quenched	1	34.1	2.75	1.79
	2	23.1	1.97	0.96
	3	11.0	1.30	0.68

CAPS 5   (CARBON-14 + ESR)

		% Efficiency	ESR	% S.D.
Air Quenched		88.0	0.227	1.46
Chemical Quenched	1	51.7	0.996	0.53
	2	17.4	1.490	0.52
Colour Quenched	1	58.5	0.608	0.84
	2	29.7	1.014	0.96

ESR = External Standard Ratio
n   = 5

belt. The standard deviations are small and similar in magnitude. Thus the geometry of the vials used and their positioning in the detector have little or no effect on the external standard ratio. In a previous study using a different model of spectrometer and different external standard the present author reached the conclusion that the vial and its position in the detector played a small but significant role in determining the ESR (11).

It is sometimes held that high sample activities cause perturbations in the ESR presumably due to pulse pile up in the analyzers. Table II shows the results of a study designed to investigate this problem. As can be seen larger standard deviations are recorded for high activity samples. Those samples which are quenched show a diminishing percentage standard deviation and this is consistant with the hypothesis that the pulse pile up is occurring. Air quenched samples containing 10^5 DPM had a S.D. = 0.64% for tritium and 0.82% for carbon-14. Further work would be necessary to ascertain the threshold counting rate at which the S.D. of the ESR starts to increase. However it is worth noting that for the maximum error observed the S.D. = 2.84% and this is not very significant in real terms since it represents only about one percent change in the counting efficiency and for most purposes this could be neglected.

CONCLUSION

In many applications and especially the double label situation external standardization offers the only convenient and accurate means of assessing counting efficiency. To the author's knowledge there is no standard published procedure to which a beginner can refer to obtain the details necessary for drawing up an ESR-efficiency curve. While the method described herein is not put forward as a protocol it is hoped that it will act as a nidus for the establishment of standard protocols in liquid scintillation spectrometry.

The results of the investigation indicate that the reproducibility of the external standard ratio for the instrument investigated is very adequate for the great majority of samples. In addition for tritium samples the ESR-efficiency curves are reasonably independent of volume as it is for chemically quenched carbon-14 samples. It is imperative that individual curves be drawn up for particular volumes of colour quenched carbon-14 samples if sample sizes are to be varied.

ACKNOWLEDGEMENTS

I thank Mrs. Margot Piekert for her competant technical assistance and for preparing the diagrams.

REFERENCES

1. E. Rapkin in The Current Status of Liquid Scintillation Counting, p. 45 (E.D. Bransome, Ed.). New York : Grune & Stratton (1970).

2. E. Rapkin in Liquid Scintillation Counting, Vol. 2, p. 61 (M.A. Crook, P. Johnson and B. Scales, Eds.). London : Heyden & Son (1972).

3. M.P. Neary and A.L. Budd in The Current Status of Liquid Scintillation Counting, p. 273 (E.D. Bransome, Ed.). New York : Grune & Stratton (1970).

4. P.E. Stanley and P.J. Malcolm to be published in Liquid Scintillation Counting, Vol. 4, London : Heyden & Son (in press).

5. P.J. Malcolm and P.E. Stanley to be published in Liquid Scintillation Counting, Vol. 4, London : Heyden & Son (in press).

6. P.J. Malcolm and P.E. Stanley, Int. J. Appl. Rad. Isotop. (in press).

7. P.J. Malcolm and P.E. Stanley, Int. J. Appl. Rad. Isotop. (in press).

8. B. Laney in Organic Scintillators and Liquid Scintillation Counting, p. 991 (D.L. Horrocks and C.-T. Peng, Eds.). New York : Academic Press (1971).

9. B. Laney in Liquid Scintillation Counting : Recent Developments, p. 455 (P.E. Stanley and B.A. Scoggins, Eds.). New York : Academic Press (1974).

10. C. Ediss, A.A. Noujaim and L.I. Wiebe in Liquid Scintillation Counting : Recent Developments, p. 91 (P.E. Stanley and B.A. Scoggins, Eds.). New York : Academic Press (1974).

11. P.E. Stanley in Liquid Scintillation Counting Vol. 2, p. 285 (M.A. Crook, P. Johnson and B. Scales, Eds.). London : Heyden & Son (1972).

STUDY OF THE SIZES AND DISTRIBUTIONS OF

COLLOIDAL WATER IN WATER-EMULSIFIER-TOLUENE SYSTEMS

by

Donald L. Horrocks
Scientific Instruments Division
Beckman Instruments, Inc.
Irvine, California 92713

The number of applications of liquid scintillation counting to the measure of radionuclides in aqueous samples has increased at a very rapid rate over the last few years. In the majority of these applications, the aqueous samples are dispersed into liquid scintillation solutions by the use of emulsifiers. The aqueous phase is dispersed as tiny micelles of much less than a few microns in size. The sizes, numbers, and distributions of these micelles could effect the real and apparent efficiency of the liquid scintillation processes. For very low energy radiations, the ability to even detect the radiations may be greatly dependent upon these properties of the colloidal systems.

In this work several radionuclides with different modes of decay have been employed in an effort to gain insight into the nature of these emulsion systems. Two different emulsion systems were studied because each showed certain characteristics which were important in the understanding of their applications to the problems of liquid scintillation counting.

Experimental

Two emulsifier systems were used in this study; Triton X-100 (Rohm and Haas) and BBS-3 (Beckman Instruments, Inc.). Triton X-100 is an alkylphenoxy-polyethoxyethanol and BBS-3 is a mixture of special purpose emulsifiers. As will be seen later, the properties of these two emulsifier systems showed quite different (not always understood) properties.

These emulsifier systems were studied using the radionuclides ^{3}H, ^{109}Cd-^{109m}Ag and ^{233}U. The ^{3}H studies supplied data for the excitations by low energy beta particles (E_{max} = 18.6 keV). The ^{109}Cd-^{109m}Ag studies supplied data for excitations by low energy conversion electrons (88 and 69 keV) and Auger electrons (29 keV). Finally, the ^{233}U studies supplied data for excitations by mono-energetic alpha particles (4.8 MeV).

All counting data was obtained using a Beckman LS-300 which
was coupled with a Nuclear Data (Model 1100) multi-channel
analyzer[1]. The pulse height spectra were obtained from
the logarithmic output pulses of the LS-300 and all spectra
show plots of the differential count rate (counts/unit
time/unit pulse height) vs. the pulse height response (channel number - 0 to 256).

In every sample the total volume of aqueous and liquid
scintillation solution was maintained at a constant volume
(12 ml) to eliminate any effects that could be caused by
difference in the geometry between the light source (liquid
scintillation solution) and the detectors (the multiplier
phototubes).

All counting was done in one of two liquid scintillation
solutions:

Solution 1 - 84% by volume Toluene
             16% by volume BBS-3

             with solutes: butyl-PBD 8g/l
                            PBBO   0.5 g/l

Solution 2 - 67% by volume Toluene
             33% by volume Triton X-100

             with solutes: PPO   5 g/l
                            M_2-POPOP 0.2 g/l

These two solutions[2,3] were chosen because they are the recommended compositions for counting tritiated water samples.

Tritium Counting

Figure 1 shows the counting efficiencies for tritium in
Solution 1 and Solution 2 as a function of the ml of H_2O
containing a constant known amount of tritiated-water.
Solution 1 (BBS-3) showed constant counting efficiency up to
a volume of 1.85 ml of H_2O at which point further H_2O
additions caused a separation of two phases with marked
decrease in counting efficiency. Solution 2 (Triton X-100)
showed a gradual decreasing counting efficiency with increasing water content. The dotted line is necessary to
indicate that for some part of the plot there was also a
separation of the system into two phases. However, at high
water content, the Triton X-100 system formed a gel which

118

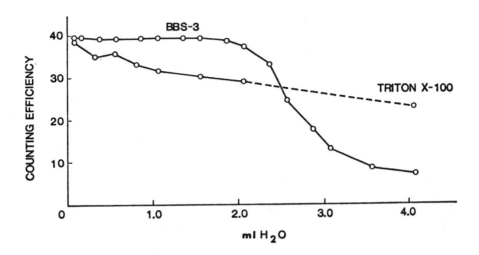

Figure 1.   Tritium counting efficiency as function of volume of ^{3}H-water in liquid scintillator solutions containing the detergent systems, BBS-3 (Solution 1) and Triton X-100 (Solution 2).

prevented the two phase from separating.

Figures 2 and 3 show spectra tritium in Solution 1 as a function of water content. Figure 2 shows the spectra for ^{3}H-toluene. The increasing water content had no effect upon the spectra. This demonstrates that the water and scintillating phases of the total water-emulsifier-toluene system are separate and the water content does not interfere with the basic scintillation processes which are taking place in the organic phase of the system. Figure 3 shows the spectra for tritiated-water and ^{3}H-toluene. The tritiated-water spectra show only a very slight shift to lower pulse heights with increasing water content. The tritiated water spectra are all shifted relative to the ^{3}H-toluene spectrum. These data show the effect of decreased pulse height as a result of the beta particles having to travel at least for a fraction of their range in the non-scintillating media, water.

Figure 4 shows three methods of monitoring quench in liquid scintillator systems as applied to the Solution 1 as a function of water content. All three methods fail in the region of two phases.

^{109}Cd-^{109m}Ag Counting

Figure 5 shows the relative scintillation efficiency for the 88 keV conversion electrons of ^{109}Cd-^{109m}Ag aqueous sample in Solution 1 and Solution 2 as a function of water content. The similarity between these plots and Figure 1 are very evident. In both cases (tritium and ^{109}Cd-^{109m}Ag), the excitations are produced by electrons.

Figures 6 and 7 show the pulse height spectra for ^{109}Cd-^{109m}Ag. Figure 6 gives the spectra obtained in Solution 1 (BBS-3) and the spectra (offset for clarity) all overlap indicating no measurable quench caused by increasing water content. Figure 7 shows the spectra obtained in Solution 2 (Triton X-100) and clearly increases in water content show dramatic shifts in the pulse height spectra similar to those which were obtained by adding a chemical quenching agent such as nitromethane. Table 1 lists data for ^{109}Cd-^{109m}Ag counted in Solution 2.

^{233}U Counting

Figure 8 shows the relative scintillation efficiencies and energy resolution for ^{233}U aqueous samples in Solution 1

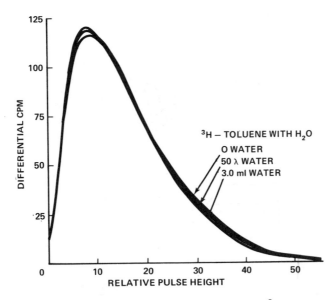

Figure 2.   Pulse height spectra for ^{3}H-toluene in Solution 1 as a function of H_2O volume.

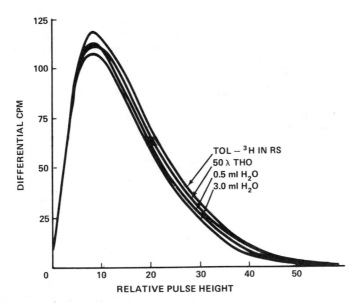

Figure 3.   Pulse height spectra for ^{3}H-toluene and ^{3}H-water in Solution 1 as a function of H_2O volume.

121

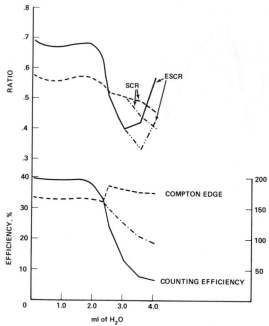

Figure 4.   External Standard Channels Ratio (ESCR), Sample Channels Ratio (SCR) and Compton Edge pulse height values as a function of the volume of water in Solution 1. Also showing the ^{3}H counting efficiency as a function of H_2O volume.

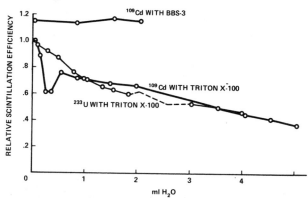

Figure 5.   Relative scintillation efficiency as a function of volume of water in liquid scintillation solutions containing BBS-3 (Solution 1) and Triton X-100 (Solution 2) for excitation with ^{109}Cd-^{109m}Ag and ^{233}U.

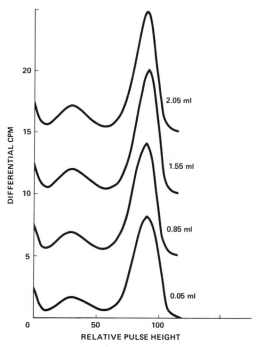

Figure 6.   Pulse height spectra (displaced on Y axis for clarity) for $^{109}Cd-^{109m}Ag$ source in Solution 1 for different volumes of work.

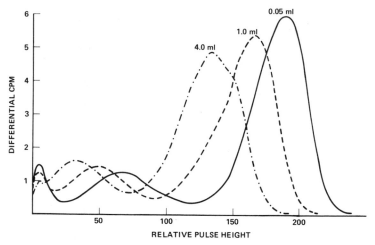

Figure 7.   Pulse height spectra for $^{109}Cd-^{109m}Ag$ source in Solution 2 for different volumes of water.

123

TABLE 1

Water Content ml	Relative Scintillation Efficiency	Counts in 88-69 keV Peak x 10^{-5} (a)
0.05	1.000	2.98
0.15	.974	2.97
0.35	.923	2.80
0.55	.875	2.88
0.85	.765	2.81
1.05	.715	2.87
1.55	.678	2.89
2.05	.660	2.90
4.05	.466	2.80

(a) Small variations most likely due to pipetting errors.

124

TABLE 2

Vol. of H_2O ml	CPM In Peak	In Tail	Total	Relative Peak CPM, %	Peak Pulse Height	Relative Scintillation Eff, %	Energy Resolution %
0.15	39,336	4,454	43,790	89.8	82.5	100.0	47.5
0.25	39,734	4,725	44,459	90.7	81.5	94.9	50.1
0.35	39,880	4,552	44,432	91.0	81.5	94.9	51.8
0.55	39,842	4,554	44,396	90.9	81.0	92.3	49.9
0.85	39,659	4,750	44,409	90.5	80.5	89.9	48.1
1.05	41,138	5,438	46,576	93.9	80.5	89.9	48.1
1.35	42,723	4,453	47,176	97.5	80.0	87.5	43.2
1.55	43,079	4,436	47,515	98.3	80.5	89.9	42.1
1.85	43,429	4,082	47,511	99.1	81.5	94.9	39.5
2.05	43,746	3,609	47,355	99.8	79.5	85.2	44.4
2.35	43,269	4,161	47,430	98.7	80.0	87.5	46.6
2.55	43,821	3,895	47,716	100.0	79.0	83.3	43.2

(BBS-3) and Solution 2 (Triton X-100) as a function of the
water content. Table 2 lists the actual data for the ^{233}U
samples in Solution 1. Figure 9 shows typical pulse height
spectra for ^{233}U in Solution 1 at two different water content
values; 1.8 ml and 2.0 ml.

Again, Figure 8 shows a close correlation with the scintillation efficiency plots of Figure 5 and the counting efficiency plots of Figure 1. The energy resolutions are approximately the same in either Solution 1 or Solution 2. Further-
more, the very high specific ionization of alpha particles
(from ^{233}U) would lead to considerable energy loss of the
alpha particles leaving the water micelles if the size of
the water micelles were increased even a small amount. This
would be reflected in a marked increase in the energy resolution. It can be seen, especially with Solution 2, that even
though the scintillation efficiency decreased, the energy
resolution stayed fairly constant. This seems to indicate
that the micelle sizes do not change with increased water
content. Rather the number of fairly uniform size micelles
has been increased.

Low Water Content Second Phase

The dip in the plot for Solution 2 in Figure 8 at low water
content, 0.1 to 0.3 ml, is due to a separation of the water
into a second phase. Figure 10 shows spectra for ^{233}U in
Solution 2. Plot (a) is the spectra for a sample containing
0.2 ml aqueous that has separated completely. Plot (b) is
the same sample which has been shaken to disperse the aqueous
phase. Even shaking the sample does not seem to completely
dispense the aqueous phase as evident by the high count rate
in the tail (below the peak) and the very wide peak. Addition of an extra 1.0 ml of water to the sample narrowed the
peak of the spectrum, eliminated most pulses from the tail
and increased the number of counts in the peak. Further,
the added water stabilized the emulsion and prevented subsequent phase separation.

High Water Content Second Phase

At water contents greater than 2.0 ml in Solution 1, the
counting system separates into two phases, see Figure 1.
Figure 11 shows spectra for a tritiated water sample of 3 ml
in 9 ml of Solution 1. Curve (a) is the spectrum obtained
when the two phases are together in a single counting vial.
The top and bottom phases were each 6 ml in volume. The two

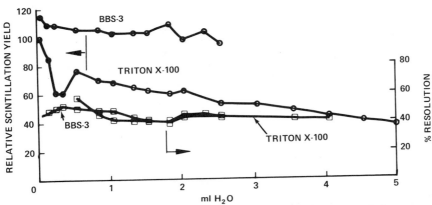

Figure 8.   Relative scintillation efficiency and %
resolution as a function of volume of water
in liquid scintillator solutions containing
BBS-3 (Solution 1) and Triton X-100 (Solution
2) for excitations with alpha particles from
^{233}U.

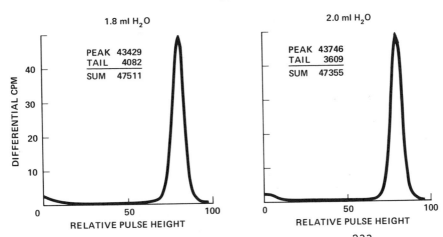

Figure 9.   Typical pulse height spectra for ^{233}U alpha
particles in Solution 1 at different volumes
of water. "Peak" denotes the counts which
are present in the peak of the distribution.
"Tail" denotes the counts which are present
at pulse heights below the peak of the distribution. "Sum" denotes the sum of counts in
the peak and tail.

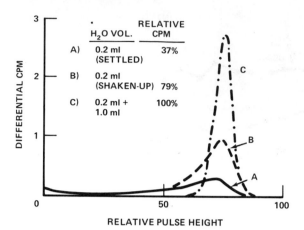

Figure 10.   Pulse height spectra for sample of ^{233}U in Solution 2 showing the stabilizing effect of added water.

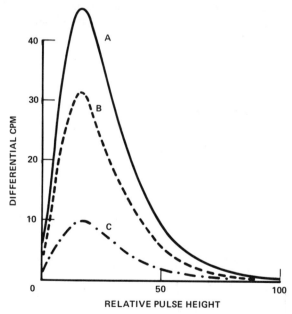

Figure 11.   Pulse height spectra for ^{3}H-water in a sample made of 3.0 ml of water and 9.0 ml of Solution 1. Curve A is spectrum obtained for two phases together in same vial. Curve B is spectrum of bottom phase alone in a separate vial. Curve C is spectra of top phase alone in a separate vial.

TABLE 3

Combined Phases

Composition		^{3}H Counting Eff., %	Phase	Volume	Compton Edge Pulse Height	% H_2 (a)
H_2O (THO)	3.0 ml		Top	6.0 ml	177	12.2
Solution 1	9.0 ml	13.7	Bottom	6.0 ml	123	37.8

Separated Phases

Composition		Compton Edge Pulse Height	CPM	^{3}H Counting Eff., %	% of Initial H_2O	Volume of H_2O
Diluted Top Phase + Solution 1	6.0 ml / 6.0 ml	169.0	16,076	40.0	24.4	0.73 ml
Bottom Phase + Solution 1	6.0 ml / 6.0 ml	160.5	39,950	32.0	75.6	2.27 ml

(a) Based on initial volume of phase and volume of H_2O in that phase.

129

TABLE 4

	Composition		^{3}H Counting Eff., %	Phase	Volume	Compton Edge Pulse Height	% H$_2$ (a)
Combined	H$_2$O (THO)	3.5 ml	8.2	Top	7.0 ml	177	16.9
Phases	Solution 1	8.5 ml		Bottom	5.0 ml	110	46.3

	Composition		Compton Edge Pulse Height	CPM	^{3}H Counting Eff., %	% of Initial H$_2$O	Volume of H$_2$O
Separated Diluted	Top Phase + Solution 1	7.0 ml 7.0 ml	169.0	22,325	40.0	33.8	1.18 ml
	Bottom Phase + Solution 1	5.0 ml 7.5 ml	160.5	35,360	32.0	66.2	2.32 ml

(a) Based on initial volume of phase and volume of H$_2$O in that phase.

phases were separated (as efficiently as possible) by drawing off the top phase which is then transferred to a second
counting vial leaving the bottom phase (with a little of the
top phase) in the original counting vial. Curve (b) is the
spectrum of the separated bottom phase and Curve (c) is the
spectrum of the separated top phase. Figure 12 shows the
Compton spectra for $^{137}Cs-^{137m}Ba$ gamma ray source; Curve (a)
both phases together, Curve (b) the bottom phase alone and
Curve (c) the top phase along. The two distinct Compton
edges indicate that each phase scintillates with its own
efficiency.

Each separated phase was diluted with extra Solution 1 to
reduce the quenching level to a measurable amount. Figure
13 shows the spectra for the Compton edge and the tritiated
water for the 6 ml of the bottom phase with an additional 6
ml of Solution 1. The Compton edge showed only a single
value which indicates the whole sample is counting with one
efficiency. The value of Compton edge pulse height indicates
the diluted solution has a 32% tritium counting efficiency.
Figure 14 shows similar spectra for the 6 ml top phase with
an additional 6 ml of Solution 1. Again, the Compton edge
showed a single efficiency which corresponded to 40% for
tritium. Tables 3 and 4 list the counting data for the
combined and separated-diluted phase and the percentage of
each phase which is H_2O.

Conclusions

These data seem to lead to several conclusions. The decrease
in counting efficiency and scintillation efficiency with
increasing H_2O content, expecially with Triton X-100, is not
due to an increase in the size of the aqueous micelles. It
is probably due to the water releasing some quencher, which
was initially bound to the Triton X-100, into the organic
phase. When the emulsion phase breaks down into two distinct
phases, each phase has scintillation character and the
radionuclides in each phase contribute to the total measured
response. Solution 2 showed two regions of phase separation;
at very low water content (0.8 - 3.3% by volume) and at
moderate water content (15 - 25% by volume) in the 12 ml
total volume samples. In cases of low water content in
Solution 2, addition of extra water to increase content to
approximately 15% by volume produced a more stable scintillation counting system without changing the scintillation
efficiency.

131

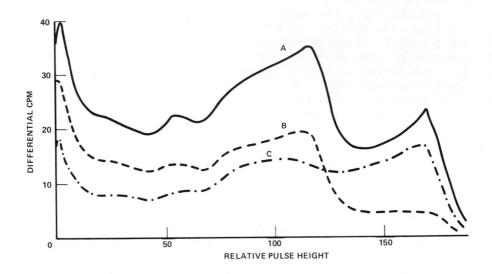

Figure 12.   Pulse height spectra for ^{137}Cs-^{137}Ba produced
Compton electrons and ^{3}H-water for same
samples as shown in Figure 11.

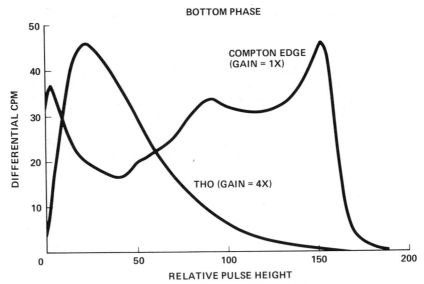

Figure 13.   Pulse height spectra for ^{137}Cs-^{137}Ba produced
Compton electrons and ^{3}H-water for bottom
phase after dilution with added Solution 1.

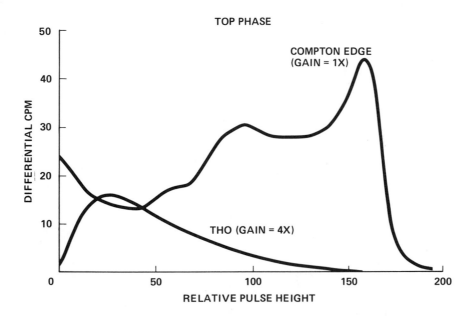

Figure 14.   Pulse height spectra for ^{137}Cs-^{137m}Ba produced
Compton electrons and ^{3}H-water for top phase
after dilution with added Solution 1.

References

1. D.L. Horrocks, <u>Nucl</u>. <u>Instr</u>. <u>Meth</u>. <u>117</u>, 589 (1973).

2. Beckman Instruments, Inc., Liquid Scintillation
    Solution Ready-Solv HP.

3. R. Liberman and A.A. Maghissi, <u>Int</u>. <u>J</u>. <u>Appl</u>. <u>Radiat</u>.
    <u>Inst</u>. <u>21</u>, 319 (1969).

EXTERNAL STANDARD METHOD OF QUENCH CORRECTION: ADVANCED TECHNIQUES

B. H. Laney
Searle Analytic, Inc.
Des Plaines, Illinois

ABSTRACT

The counting efficiency for a liquid scintillation sample varies with its composition. Impurities from solubilizers as well as the specimen itself typically quench the available energy, reducing both the pulse heights and the total number of detectable events. Counting efficiency as measured in a pulse height channel is therefore uncertain, particularly when assaying low-energy radionuclides or dual labeled samples.

An external source of gamma radiation is frequently utilized to quantitate the degree of quenching. This is done by analyzing the additional pulse height spectrum created from Compton electrons, which are produced in the sample by the radiation from the gamma source. The resulting data may then be used to correct for losses in pulse height and counting efficiency. The counts induced by the gamma source are usually measured in two pulse height channels and the ratio of the counts in these channels is used to quantitate quenching. This is called the external standard ratio (ESR). The fixed windows in these channels, however, limit the useful dynamic range of quench determination or may produce unacceptable statistical uncertainty in the estimated counting efficiencies of quenched samples.

A new method of quench determination based upon relative pulse height (RPH) is described. Applying this technique to the external standard Compton spectrum provides a means for quantitating quench in terms of RPH over a 30 to 1 change with less than $1\frac{1}{2}\%$ standard deviation. By recognizing the quantum characteristic of weak scintillations, accurate pulse height restoration can be extended to quenched samples containing low energy radionuclides such as ^{3}H and ^{14}C. Isotope energy, window settings, and true disintegration rate can be computed from the RPH quench value of unknown samples without empirical curves.

INTRODUCTION

Most commonly used methods for quantitating quench in a homogeneous liquid scintillation sample employ the use of an external gamma ray source. Each Compton electron induced in the sample by the penetrating gamma radiation transfers its energy to the solute in the same manner as an electron ejected from the nucleus of an atom undergoing beta decay. The net Compton spectrum is independent of the activity, energy, number and type of radionuclides that may be present in the sample. This universality largely accounts for the appeal of the method.

Gamma emitting radionuclides have been used as early as 1950 to measure the relative scintillation efficiency of organic solutes and solvents in liquid scintillation samples. Reynolds et al.[1] used the coincidence/non-coincidence ratio of Compton electrons induced in the test sample by a ^{60}Co source to quantitate scintillation efficiency. Kallman[2] used the relative pulse height (RPH) as measured by the peak intensity from a radium source to quantitate relative scintillation efficiency.

The Compton spectrum induced in the liquid scintillation sample by a gamma ray source has also been correlated with the counting efficiency of beta emitting radionuclides. Kaufman et al.[3] combined the external standard technique with the sample channels ratio technique described by Baille[4] to correlate ^{57}Co channels ratio with ^{3}H efficiency. Fleishman et al.[5] correlated the count rate induced by a ^{60}Co source with the counting efficiency of ^{40}K. Similarly, Higashimura[6] correlated the count rate from a ^{137}Cs gamma source with ^{14}C counting efficiency. Horrocks[7] correlated the RPH of a ^{137}Cs Compton edge as measured by its pulse height at half maximum intensity with ^{14}C counting efficiency.

Quench measurements using either the ESR or RPH methods are unaffected by the half-life of the gamma source and also tend to produce congruent efficiency vs quench curves for samples differing in sample volume and electron density[7].

Two advantages of the RPH method over the ESR method are 1) the useful range of quench is not bounded by fixed windows and 2) the RPH has physical significance for determining counting efficiency and quench corrected window settings[8],[9],[10].

When the spectrum shift due to quenching is extensive, part of the spectrum from one radionuclide may fall outside its fixed counting channel and may spill into another channel set to count another radionuclide. Two methods for automatically adjusting the counting windows for quenched samples have been investigated by Wang[11] and Herberg[12]. The objective of the method described by Wang, automatic quench correction (AQC), is to increase the electronic amplification to restore the sample's pulse height spectrum to the level it would have had were it not quenched. The degree of quenching is measured by the ESR. This method has two inherent difficulties. Wang demonstrated that separate correction relationships are required to restore the end points of ^{3}H and ^{14}C. Although the external standard windows can be tailored to approximate a linear response over a limited range of quench, the method is not linear over a wide range of quench. Secondly, although a single gain correction function may be adequate for moderately quenched samples containing high energy isotopes, it is inadequate for low energy isotopes or heavily quenched samples. Consequently, separate gain adjustments are required for different isotopes and clearly a single gain adjustment is not valid for all isotopes.

The automatic activity analyzer (AAA) described by Herberg uses ten sets of individually adjustable preset windows for each isotope corresponding to ten preset levels of quench. By electronically quenching each unknown sample to one of the preset levels of quench, the sample is adjusted to the preset window. This method creates three problems not present in the AQC system:

1) hours of operator time and skill are required to properly standardize the system for each isotope and counting condition;

2) each sample must be exactly quenched to one of the ten preset levels which takes in excess of three minutes;

3) since the sample is quenched further by the instrument prior to counting, it will be counted at less than the available efficiency, somewhat defeating the main objective of optimizing the settings to improve counting statistics.

A major advantage of the AAA method is that disintegration rate is obtained directly without curve fitting.

This paper describes a method for computing and setting pulse height analysis channels for any combination of isotopes at any level of quench without additional operator or machine time. Window levels are computed from mathematical relationships, RPH and the unquenched window settings.

THEORY

RPH has been used to quantitate relative scintillation efficiency by many investigators[1,2,7,8,9,13]. However, relative scintillation efficiency, the relative number of photons generated per keV expended, is not always proportional to RPH because:

1) the number of photons emitted per keV of energy depends upon the energy of the primary beta particle[10];

2) the emission spectra of the two solutions being compared may have different optical response at the photo-detector[13].

3) the pulse height at the output of a phototube is not proportional to the number of impinging photons for weak scintillations.

The latter case can be demonstrated by comparing the pulse height of a large 1000 photon scintillation with that from a weak single photon scintillation. Assuming a photocathode quantum efficiency of 0.30, the pulse height from each 1000 photon scintillation is statistically equal to 300 photoelectrons. However, the pulse height from a single photon scintillation, when detected, is equal to the pulse height of a single photoelectron. Although the relative photon input for the two cases is 1000 to 1, the relative pulse height output is only 300 to 1. The difficulty arises because of the finite light quanta and the less than 100% probability of detection. A single photon has a relative pulse height 1/0.30 greater than would be required for proportionality. However, since it is only detected 30% of the time, the charge delivered from 1000 single photon scintillations is statistically equal to that from a single 1000 photon scintillations. Therefore, on the average, the number of photoelectrons, but not necessarily the pulse height, is proportional to the number of photons per scintillation.

Pulse height shift with quenching are compared for ^{133}Ba Compton, ^{14}C and ^{3}H spectra in figures 1 A, B, and C respectively. It can be readily seen that the pulse height shift for ^{3}H is much less than for ^{133}Ba and ^{14}C. For spectra below about channel 10, count rate declines faster and pulse height shifts less rapidly with quenching due to the quantum threshold. This data compares with that of Wang[11].

When analyzing relative pulse heights less than about 10 photoelectrons per scintillation, it is necessary to account for the quantum threshold and the dependence of pulse height upon the number of photoelectrons detected. Multiple phototube coincidence detectors further alter the pulse height distribution because the pulse height at one photo-tube is dependent upon both detection probability and pulse height from the second phototube. The method of combining the pulse heights from the detectors typically alters the pulse height distribution as a function of the number of photoelectrons[14].

Since detection efficiency ultimately depends upon the number of photoelectrons liberated from the phototransducer, relative pulse height corrected for the quantum threshold can be used to monitor changes in overall system efficiency. The calculated relative pulse height, RPH_i of a spectrum with measured average lesser pulse height PH_i, relative to a reference spectrum with average lesser pulse height PH_R, corrected for the single photon intercept, is approximated by:

$$1) \quad RPH_i = \frac{PH_i - PE_R}{PH_R - PE_R}$$

Some of the methods for quantitating the pulse height of a spectrum include: the channels ratio[3,4], the pulse height at the peak intensity[2], the pulse height of the upper edge[7], and the average pulse height.

The average pulse height of the Compton lesser spectrum from ^{133}Ba was selected as the method for quantitating RPH for the following reasons:

1. Relative pulse height can be measured directly without empirical relationships.

2. High statistical precision is obtained from the average because of the high count rate obtained using all, rather than part of, the spectrum.

3. The average pulse height of the lesser spectrum is a good analog of the average efficiency of detection over a spectrum of pulse heights which may be distorted due to energy dependent quench mechanisms.

4. ^{133}Ba was selected for its relatively low energy which more closely approximates that of low energy beta emitters.

An ideal external standard gamma source for assaying beta emitting radionuclides with maximum energy less than 200 keV should have a Compton electron spectrum extending to about 200 keV: high enough to track large changes in quench, low enough to track energy dependent quench mechanism and low enough so as not to produce instability from phototube fatigue. Ideally, the Compton spectrum should overlap the same energy range as the isotope being assayed, so that energy dependent quenching mechanisms[10] will alter the Compton spectrum in the same manner as the beta spectrum and thus produce a change in the external standard pulse height, which is analogous to that of the sample spectrum. Under these conditions, the external standard response will tend to track energy dependent changes in efficiency from different solvent-solute-quencher systems.

Gamma sources which produce most of their Compton electrons beyond 200 keV are equally capable of monitoring changes in scintillation efficiency which are not energy dependent, but do a poor job of tracking energy dependent quenching of low energy beta emitters such as ^{3}H and ^{14}C.

High energy gamma sources also tend to cause phototube drift. Phototube fatigue occurs at high anode current. A high activity gamma source sufficient to produce 10^5 to 10^6 counts per minute is normally employed to achieve a low statistical uncertainty in the quench measurement in less than one minute. The higher the energy count rate product of the external standard source, the greater the anode current and consequently phototube amplification is more unstable. It is therefore important to use a gamma source with energy sufficiently high to produce a statistically significant pulse height shift over wide ranges of

140

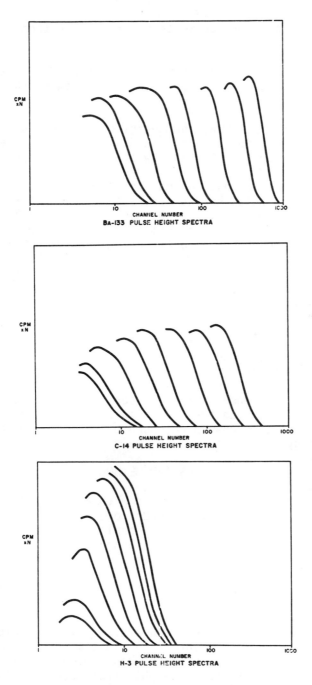

Figure 1   Pulse Height Spectra of ^{133}Ba, ^{14}C and ^{3}H

quench, yet not so energetic as to produce instability in the measurement.

Thus, ^{133}Ba with a peak Compton energy of 207 is keV is an almost ideal gamma source for assaying low energy beta radionuclides.

Once the RPH is known, the energy of the beta spectrum can be estimated and the absolute disintegration rate DPM of the sample can be determined without quench curves.

The procedure for estimating the isotope energy is as follows: 1) the pulse height of the sample is accurately measured by accumulating 200,000 counts, 2) the pulse height of the sample spectrum is quantitated by computing the average pulse height. The pulse height energy, PHE_i, of the sample is computed from the measured pulse height of the sample, PH_i, and the RPH determined with the external standard as follows:

2)
$$PHE_i = \frac{PH_i - PE}{RPH_i} + PE$$

PHE_i is the estimated pulse height of the sample at the reference quench at a RPH = 1. PE is the average pulse height intercept. It has a value of 1.35 channels as determined from quench ^{3}H with lesser pulse height analysis. Beta energy in keV can then be calculated using the correction for ionization quenching shown by Gibson[10] and a single energy calibration.

MATERIALS AND METHODS

All measurements were performed with a Searle Analytic Mark III liquid scintillation spectrometer. Nine microcuries of ^{133}Ba hermetically sealed in a stainles steel pellet is mechanically positioned beneath the center of the sample. EMI type 9750QB phototubes are operated with linear dynode voltage distribution at a nominal voltage of 1300 volts (a block diagram of the spectrometer is shown in Figure 2). The pulse heights from each of the phototubes are simultaneously applied to the coincidence circuit and two 12-bit analog to digital convertors. The coincidence circuit, through the bus interface, initiates the simultaneous conversions from both phototubes. Upon completion of the digitizing of the pulse heights from both phototubes, the numerical value corresponding to the linear pulse height from each of the phototubes is transmitted to a micro-

142

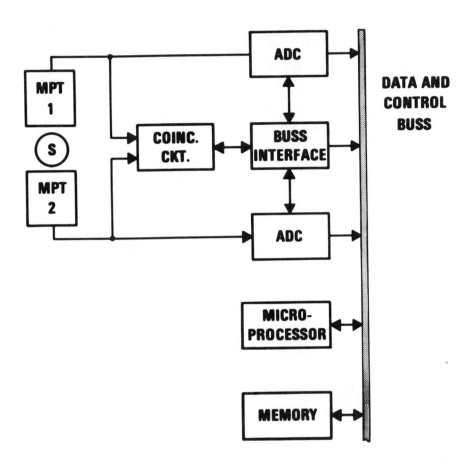

Figure 2   Pulse Height Analyzer Block Diagram

processor via the data bus. A test is made to insure that
the two pulse heights are properly correlated[15] and then the
smaller of the two numerical numbers corresponding to the
lesser pulse height[14] is then analyzed and stored.

In one mode of operation, data is collected in the memory
as it would be in a multi-channel analyzer and decisions
about pulse height analysis are made subsequent to the
collection of all sample data. This mode is used for storing
the spectra of standards, and for external standardization.
In another mode, when assaying unknown samples, quench
corrected windows are numerically determined prior to sample
counting, and pulse height analysis is determined in real
time.

The procedure for determining the RPH from the external
standard, is as follows: first, the ^{133}Ba external standard
is positioned beneath the sample; the gross external
standard spectrum is accumulated in the multi-channel mode
for a period of 0.4 minutes. Then, the external standard is
removed to its shielded storage location and the sample
spectrum is accumulated for 0.4 minutes and subtracted from
the gross external standard spectrum. Finally, RPH is
computed from the net external standard spectrum.

3)
$$RPH_i = \frac{ESPH_i - ESPH_\infty}{ESPH_R - ESPH_\infty} = \{ESP\}^{-1}$$

The reference external standard pulse height, $ESPH_R$, is a
constant determined from a measurement of the external
standard pulse height with a sealed nitrogen flushed toluene
standard using PPO and POPOP as primary and secondary fluors.
This constant has a value of 135 channels and varies from
one instrument to another within a range of ± 8 channels.
The external standard pulse height at infinite quench,
($ESPH_\infty$) is a constant having a value of 1.75 channels. $ESPH_i$
is the external standard pulse height measured on any
unknown sample i. RPH_i is the true relative pulse height
for sample i relative to the reference standard, corrected
for the single photoelectron intercept. The inverse of
the relative pulse height is defined as the ESP. Efficiency
vs quench curves are typically more linear when plotted vs.
the inverse relative pulse height, ESP.

Quench corrected window settings are computed from their
unquenched value L_R and the relative pulse height of
individual samples as follows (Figure 3): all lower levels

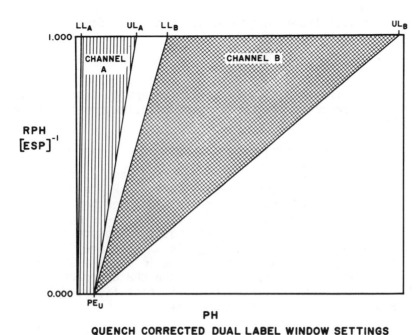

Figure 3 Quench Corrected Dual Label Window Settings

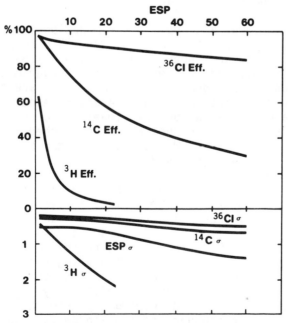

Figure 4 Efficiency And Standard Deviation vs ESP

except for dual label, high energy isotope windows are computed as follows:

4) $$L_i = (RPH_i) L_R$$

All upper levels and dual label high energy isotope lower levels are corrected for the single photoelectron intercept as follows:

5) $$L_i = (RPH_i)(L_R - PE_U) + PE_U$$

L_R is any level setting for an unquenched sample, i.e., when the relative pulse height is equal to 1.0. $PE_U = 4.70$ channels, determined from measurements of quenched ^{3}H. Since all pulse height information is lost below the single photo-electron level, discriminator settings for isotope separation below this level are meaningless. Therefore, the lower level of the high energy isotope in a dual-labeled experiment approaches this level as the RPH approaches zero. In this way, only statistically significant data is maintained in the high energy channel.

For single labeled isotopes, the lower level and upper level discriminators converge to zero and PE_U respectively as the relative pulse height approaches zero.

Unquenched ^{3}H lower and upper discrimination levels in program 8, channel B are typically 0.8 and 25.0 respectively. Unquenched ^{14}C lower and upper levels in program 8, channel B are typically 10.1 and 321 respectively. Program 9 window settings for unquenched dual labeled ^{3}H and ^{14}C are typically 0.8 to 9.1 and 25 to 321 in channels A and B respectively. Program 1, channel B has unquenched levels to 0 to 26 for assaying ^{3}H. Program 2, channel B has unquenched levels of 0 to 315 for assaying ^{14}C. Window settings may be computed at any RPH using equations 4 and 5.

The method for determination of absolute disintegration rate (DPM) is as follows: the instrument is standardized for a given isotope by counting a series of quenched standards and the total multi-channel spectrum of each standard is permanently stored along with the known DPM. To assay an unknown, the RPH is determined using the external standard, the spectrum of the unknown isotope is computed from the stored spectra at that RPH, efficiency is computed from the constructed standard spectrum by integrating the count rate within the boundary of the window used to assay the unknown sample, and the calculated efficiency is divided

TABLE I

SINGLE LABEL ^{14}C EFFICIENCY
WITH QUENCH CORRECTED WINDOW SETTINGS

RELATIVE PULSE HEIGHT $(ESP)^{-1}$	FIXED WINDOW EFFICIENCY %	CORRECTED WINDOW EFFICIENCY %	RELATIVE INCREASE %
0.83	87	89	2
0.26	63	84	33
0.15	43	78	81
0.111	27	72	170
0.083	15	63	320
0.050	3	42	1,400
0.037	0.5	29	5,800

TABLE II

DUAL LABEL ^{14}C EFFICIENCY
WITH QUENCH CORRECTED WINDOW SETTINGS

RELATIVE PULSE HEIGHT $(ESP)^{-1}$	FIXED WINDOW EFFICIENCY %	CORRECTED WINDOW EFFICIENCY %	RELATIVE INCREASE %
0.83	68	71	4
0.59	58	70	20
0.40	45	68	52
0.26	28	64	132
0.15	8.0	56	600
0.11	1.8	48	2,600
0.08	.1	39	39,000

into the count rate from the unknown. In other words, a standard spectra is computed at the same quench as the unknown sample and integrated in the same window used to count the unknown sample in order to determine the efficiency for counting the same isotope, at the same level of quench and in the same window as the unknown.

RESULTS

Fixed and quench corrected window settings for single labeled ^{14}C are compared in Table I. Fixed window settings in program 8 were set at RPH=0.83. The improvement in counting efficiency for the moderately quenched samples having RPH = 0.111 is 170%, whereas, the improvement in the heavily quenched sample having RPH = 0.050 is 1400%.

The improvement in counting efficiency with pulse height restoration is even more dramatic when narrower counting channels are used, as when required for assaying dual labeled samples in program 9. Table II shows the improvement when assaying ^{14}C in the presence of ^{3}H. At a RPH = 0.11, the improvement in counting efficiency is 2600%. Heavily quenched samples are uncountable in the fixed counting windows, whereas, the quenched corrected windows gave counting efficiencies in excess of 30%.

The ability to track different energies as a function of quench, was determined by counting ^{3}H and ^{14}C separately as the low energy isotope in program 9. In this program, the lower level of the B channel is set to the same value as the upper level of the single label program 8. Table III shows the excellent tracking of the upper level for both ^{3}H and ^{14}C below 6% relative scintillation efficiency for ^{3}H and below 2% relative scintillation efficiency for ^{14}C.

The precision of the DPM measurement of quenched samples was determined by computing the standard deviation (S.D.) from 30 conveyor cycles in the ^{3}H single labeled program 1 and the ^{3}H dual labeled program 5 and for ^{14}C in the single labeled ^{14}C program 2 and the dual labeled ^{14}C program 5. The S.D. is less than 1½% for both isotopes under both single and dual labeled counting conditions quenched down to a RPH of 11%. For ^{14}C, the reproducibility of the disintegration rate measurement is maintained at less than 1½% down to a RPH of 2%. For ^{14}C, the S.D. increases to about 2½% at an RPH of 7%. However, since the counting efficiency for ^{3}H at this level of quench is only 4%, this corresponds to an absolute efficiency uncertainty of only 0.1% (table IV)

The efficiencies and standard variances are shown in Figure 4. As expected, high energy isotopes can be assayed more accurately than low energy isotopes. High energy isotopes have flatter quench curves and are less effected by variations in the quench parameter. It is interesting to note that the % S.D. in the ESP value is greater than % S.D. in the disintegration rate DPM derived from the ESP as shown for ^{14}C and ^{36}Cl. Measurements of the S.D. of an external standard quench parameter alone are not very meaningful. Variations in sample geometry, vial thickness and dirt accumulation may cause changes in the counting efficiency and corresponding changes in the quench parameter. Since both measurements are correlated, a component of the variance is eliminated in the DPM calculation.

TABLE III

TRACKING OF ESP CORRECTED LEVELS

TRITIUM EFFICIENCY (%)			CARBON 14 EFFICIENCY (%)		
RPH	L TO U	U TO ∞	RPH	L TO U	U TO ∞
0.71	55	0.5	0.71	86	0.2
0.55	50	0.5	0.40	84	0.3
0.38	41	0.5	0.34	84	0.3
0.18	23	0.4	0.28	84	0.3
0.13	17	0.3	0.044	65	0.4
0.086	11	0.2	0.026	52	0.5
0.056	6.9	0.1	0.020	43	0.5

TABLE IV

PRECISION OF DPM MEASUREMENT

TRITIUM % VARIANCE			CARBON 14 % VARIANCE		
RPH $(ESP)^{-1}$	SINGLE LABEL	DUAL LABEL	RPH $(ESP)^{-1}$	SINGLE LABEL	DUAL LABEL
0.77	0.4	0.3	0.77	0.2	0.2
0.67	0.4	0.3	0.40	0.1	0.3
0.50	1.1	1.1	0.17	0.2	0.4
0.34	0.5	0.6	0.09	0.3	0.6
0.22	0.9	0.3	0.05	0.3	1.1
0.11	1.1	1.2	0.031	0.7	0.9
0.07	2.3	2.5	0.017	0.7	1.4

CONCLUSIONS

The average pulse height of the lesser spectrum from a ^{133}Ba Compton spectrum has been used to reproducibly quantitate quench in a liquid scintillation system in terms of RPH. The method is useful for quantitating quench over a 60 to 1 change in RPH. Pulse height response is non-linear for weak scintillations; the non-linearity is dependent upon whether a single phototube or two in coincidence are employed and upon how the signals from each are combined. By correcting the pulse height response to weak scintillations for the single photon intercept, RPH can be used to quantitate relative scintillation efficiency assuming the photon emission spectrum probability is constant.

Quench corrected window settings can be accurately determined from the external standard RPH for both ^{3}H and ^{14}C without operator intervention, calibration curves, loss in counting efficiency or additional time. Counting efficiency is substantially improved particularly for heavily quenched and dual labeled samples. The method of pulse height restoration should be applicable for all beta emitting isotopes.

The disintegration rate can be reproducibly determined for single or dual labeled ^{3}H and ^{14}C from the RPH of the external standard without quench curves.

The beta energy of unknown isotopes can be estimated from the measured pulse height of the sample spectrum and the external standard spectrum assuming the photon emission spectrum is constant.

BIBLIOGRAPHY

1. G. T. Reynolds, F. B. Harrison, G. Salvini, Physics Review 78,488 (1950)

2. H. Kallman, Physics Review 78, 621 (1950)

3. W. J. Kaufman, A. Nir, G. Parks, R. M. Hours in Tritium in the Physical and Biological Sciences, Vol. 1, p. 249, IAEA, Vienna (1962)

4. L. A. Baille, International Journal of Applied Radiation and Isotop-s 8, 1 (1960)

5. Fleishman, D. G., Glazunov, U. V. Pribory i Teknika Eksperimenta, 3, 55 (1962)

6. T. Higashimura, International Journal of Applied Radiation and Isotopes, 13, 308 (1962)

7. D. L. Horrocks, Nature 202, 78 (1964)

8. J. A. B. Gibson and H. J. Gale, International Journal of Applied Radiation and Isotopes, 18, 681 (1967)

9. K. F. Flynn, L. E. Glendenin and V. Prodi in Organic Scintillators and Liquid Scintillation Counting, p. 687, (D. L. Horrocks and C. T. Peng eds.) Academic Press, New York, 1971

10. J. A. B. Gibson in Liquid Scintillation Counting, Vol. 2, p. 27, (M. A. Crook, P. Johnson and B. Scales eds.) Heyden and Son, Ltd., (1971)

11. C. H. Wang in The Current Status of Liquid Scintillation Counting, p. 305, (E. D. Bransom, Jr., ed.), New York Grune and Stratton (1970)

12. R. J. Herberg in Organic Scintillators and Liquid Scintillation Counting, p. 783: (D. L. Horrocks and C. T. Peng, eds.) New York: Academic Press (1971)

13. R. K. Swank and W. I, Bucks, Review of Scientific Instruments, 29, number 4, 279 (1958)

14. B. H. Laney in Liquid Scintillation Counting, Vol 4, p. 74 (M. A. Crook, ed.) London: Heyden and Son, 1976

15. B. H. Laney in Organic Scintillators and Liquid Scintillation Counting, p. 991 (D. L. Horrocks and C. T. Peng, eds.) New York: Academic Press (1971)

LIQUID SCINTILLATION COUNTING OF NOVEL RADIONUCLIDES

J.A.B. Gibson

Environmental & Medical Sciences Division,
AERE, Harwell, Oxon, UK

ABSTRACT

The theoretical background of counting radionuclides in liquid scintillators is presented. The effects of quenching and finite scintillator size are briefly described and the theory is justified by an experimental comparison between ^{55}Fe and ^{3}H in which all facets of the theory are important. Counting efficiencies for other nuclides decaying by 100% electron capture are calculated and compared with efficiencies for the β emitters ^{3}H, ^{14}C and ^{36}Cl. Also included are comments on the special problems associated with counting plutonium in biological materials. The essential conclusion is that in order to improve the technique and avoid unnecessary pitfalls it is necessary to have a sound understanding of the underlying theory of liquid scintillation counting.

INTRODUCTION

Liquid scintillation counting is a well established method for the measurement of a very wide range of radioactive isotopes. When a new radionuclide is used in an investigation the question arises as to whether it can be measured in a liquid scintillator. At this stage a series of questions should be asked.

(i) What is the nature of its activity and is the decay scheme known so that it can be standardised and counting conditions can be optimised.

(ii) Is there enough energy available for decay events in the scintillator to be detected by the photomultiplier.

(iii) Is there a suitable compound of the isotope which can be incorporated into one of the many scintillator cocktails available, without undue quenching of the light output.

It is the answers to the first two questions that can be answered from an understanding of the basic interactions in the scintillator. The choice of scintillator cocktail is available in a wide range of reviews and conference proceedings eg (1-8).

Over the years the counting efficiency of instruments has increased and the background has been reduced as can be seen in table I(1). The counting efficiency for other isotopes emitting β^- or β^+ radiation has similarly improved and the efficiency for α emitting isotopes can be 100% in most cases. This leaves the important range of nuclides which decay by electron-capture (EC). In this decay it is X-rays and Auger electrons which must be measured in the scintillation counter and it is the measurement of these nuclides which will form the main basis of this paper.

TABLE I

Improvements in Instrument Performance(1)

	^{3}H		^{14}C	
	Counting Efficiency %	Background cpm	Counting Efficiency %	Background* cpm
1954	10	80	75	60
1962	25	55	80	30
1964	40	30	85	25
1969	60.	20	90	16
1972	65	18	97	11

*Above ^{3}H end point.

In the paper I will present a brief introduction to the underlying theory of liquid scintillation counting, showing how the counting efficiency can be calculated for unquenched and quenched solutions of any isotope and how the theory is applied to EC nuclides. Comparisons with the counting efficiencies for β emitting isotopes will be made and the special problems of plutonium counting in biological material will be mentioned.

THEORY OF THE SCINTILLATION COUNTER

The theoretical approach in this section is a summary of a series of papers (9-13) and is specifically for single electron events in the scintillator as opposed to the theory for a β spectrum. The section includes a discussion of losses in the scintillator and photomultiplier, calculation of the counting efficiency, allowance for escape of X-rays from the scintillator without interaction, and the effects of quenching.

The essentials of the scintillation counter

The conversion of the initial electron energy into light photons is a complex process as detailed by Birks (14) and the essential details are set out in table II. The first important point is that the scintillation efficiency at high electron energies is about 4% for anthracene and for liquid scintillators may be about 1% as shown in table II.

TABLE II

Energy Transfer Processes in a Liquid Scintillator

Process	Energy Transfer	Gain 5 keV	Quanta keV^{-1}		Quanta 5 keV
			1 MeV	5 keV	
Electron input	Energy deposition*		340	340	1730
Solvent Excitation	dE/dx losses at 5 keV	0.63	340	220	1080
Solute Excitation	Quenching and light losses				
Light Emission		0.01	3.4	2.2	10.8

*Assuming all the light is produced at 425 nm

Secondly at low energies, as shown by Gibson and Gale (10), following Birks (14), as the rate of energy deposition of a particle (dE/dx) increases then the light output per unit energy input is decreased ie the scintillation efficiency is reduced,

$$\frac{dL}{dE} = \frac{S(dE/dx)}{1 + kB(dE/dx)} \qquad (1)$$

Where S is the scintillation efficiency and kB is a constant. The extent of this effect can be seen in figure 1 (10) in which the relative scintillation efficiency is plotted against electron energy.

The counting efficiency

The output from the scintillator is subject to normal statistical variations and at 1 MeV the distribution is conveniently approximated by a Gaussian distribution. At 5 keV with only 10.8 photons (on average) the distribution is a Poisson of the form,

$$G(r) = \frac{m^r e^{-m}}{r!}$$

where m is the mean number of photons and G(r) is the probability of having r photons in a given event. The 'zero probability' (9) is given when $r = 0$ ie $G(0) = e^{-m}$, and for m = 10.8, $G(0) = 2.04 \times 10^{-5}$. This is small and thus if we could count every photon then the counting efficiency for a 5 keV electron in the scintillator would be essentially 100%.

However, the efficiency of conversion of photons to electrons in a photomultiplier varies from 10 to 40%. If a is the conversion efficiency from photons to electrons at the input to the first stage of the photomultiplier then the zero probability at this stage is (13)

$$Z = \exp\{-m(1-e^{-a})\} \qquad (2)$$

which approximates to $Z = e^{-am}$ when a is small. If we let $n = am$ then $Z = e^{-n}$ which is the zero probability for a Poisson distribution with a mean of n. The effect of later stages in the photomultiplier are small ($<1\%$) if the first stage gain exceeds 5. The counting efficiency is $\epsilon = 1-Z$. Using the data from table II the counting efficiency is given in table III. It should be noted that this is the maximum achievable efficiency at zero bias level on the discriminator following the amplifier and in practice lower efficiencies will be obtained.

The efficiency for two photomultiplier tubes in coincidence is slightly more complicated in that at least two photons must be produced and one photon must be detected by each photomultiplier. The distribution is of the form

$$G(r) = (1-0.5^{r-1}) \, n^r \frac{e^{-n}}{r!} \qquad (3)$$

where $n = m(1-e^{-a})$ as before. Equation (3) only applies where $r \geqslant 2$ and the counting efficiency for the two tubes in coincidence, ϵ_c is simply the product of the counting efficiencies of the single tubes (ϵ_s). If the two tubes have equal counting efficiencies then $\epsilon_c = \epsilon_s^2$ as given in table III. In order to determine the counting efficiency for a β spectrum it is necessary to integrate over the whole spectrum (13).

The use of the coincidence system greatly reduces the background but it also reduces the counting efficiency and these two effects have to be balanced in a given application.

TABLE III

Counting Efficiency for Single and Coincidence Systems for 5 keV Electrons

Photocathode Efficiency, a %	Mean no. of electrons $n = am$	Zero Probability		Efficiency %	
		e^{-n}	Equn (2)	$\epsilon_s = 1-Z$ Single	ϵ_s^2 Coincidence
10	1.08	0.340	0.358	64.2	41.2
20	2.16	0.115	0.141	85.9	73.8
30	3.24	0.039	0.061	93.9	88.2
40	4.32	0.013	0.028	97.2	94.4

Figure of merit

It is useful to specify an overall efficiency for the conversion of input of electron energy into a mean numbers of electrons at the first dynode of the photomultiplier n(E). This is defined as

$$P = \frac{n(E)}{E.F(E)} \qquad (4)$$

where F(E) is the relative scintillation efficiency as given in figure 1. P is independent of energy E and for the system described in tables II and III, $P = a \ 3.4 =$ 0.34 e keV^{-1} at a photocathode efficiency of 10%. It is possible to measure P by various methods as discussed by Gibson (13).

The effects of quenching

A major problem in liquid scintillation counting is a reduction in the scintillation efficiency due to impurities introduced with the sample. This quenching effect can take two main forms (a) chemical and (b) colour quenching.

Chemical quenching involves the interception of energy in the transfer and diffusion processes between molecules. It can be thought of in simple terms as follows

(i)   Energy absorption $M+E \rightarrow M^*$ Rate = 1

(ii) Light emission $M^* \rightarrow M+h\nu$ Rate = $k_1 [M^*]$

(iii) Quenching $M^*+Q \rightarrow$ Loss   Rate = $k_2 [M^*][Q]$

where M is a fluorescent molecule in the ground state
$\quad$ M* is a fluorescent molecule in an excited state
$\quad$ Q is a quenching molecule
$\quad$ [ ] are molecular concentrations
$\quad$ $k_{1,2}$ are relative reaction rate constants.

Then the total available fluorescence is

$$S_o = k_1[M^*] + k_2[M^*][Q]$$

and fluorescence from a quenched solution is

$$S = k_1 [M^*]$$

The relative quenching factor is then

$$g = \frac{S}{S_o} = \frac{1}{1+kw} \qquad (5)$$

where $k = k_1/k_2$ and $w = [Q]$. Equation (5) is a Stern-Volmer equation and is very similar to the equation (1) for ionisation quenching in the track of the ionising particles where kw is replaced by $kB(dE/dx)$. The factor g for chemical quenching is independent of electron energy and reduces the figure of merit from P to gP and thus

157

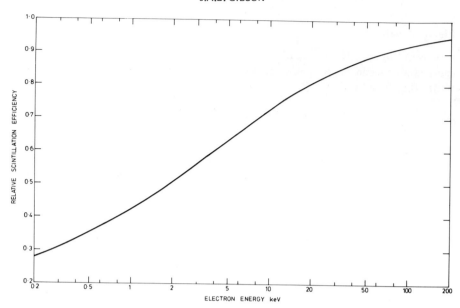

FIG.1. EFFECT OF IONISATION QUENCHING ON THE SCINTILLATION EFFICIENCY

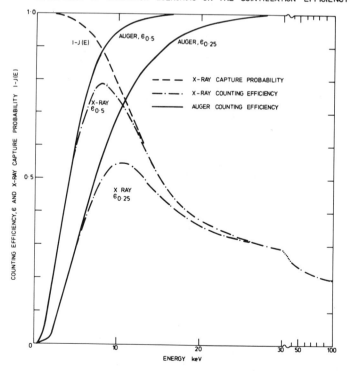

FIG.2. CALCULATED COUNTING EFFICIENCY FOR AUGER ELECTRONS
AND X-RAYS

changes $\exp(-n)$ to $\exp(-gn)$ and so reduces the counting efficiency. This is discussed more fully elsewhere (11).

Colour quenching involves an interception of the light-photons and may follow equation (5). At high quench levels it will distort the spectrum and there is no really satisfactory theory to predict the changes in counting efficiency, eg ten Haaf (15).

Loss of energy into the scintillator walls

The discussion has so far assumed that the scintillator is infinite in volume and for electrons this is a reasonable assumption for volumes of about 15 cm^3 at energies up to 100 keV. Even above this energy enough energy is deposited in the scintillator to produce a pulse in the photomultiplier and so the counting efficiency would be 100%. However for X (and γ) rays there is a high probability of escape from the scintillator without any interaction at all. This escape probability is given by $e^{-\mu d}$ where d is the distance the X-ray travels in the scintillator. Integrating $e^{-\mu d}$ over all solid angles for a cylindrical volume the total probability of escape $J(E)$ can be calculated and the capture probability $(1-J(E))$ determined (figure 2 (16)). This capture probability is multiplied by the counting efficiency for electrons to obtain a counting efficiency for X-rays in a given scintillator (figure 2).

COUNTING OF NUCLIDES DECAYING BY ELECTRON CAPTURE

Electron capture is a mode of nuclear decay in which K or L electrons are captured by the nucleus resulting in a vacancy in one of these shells. In the subsequent re-arrangement of the electrons within the shells either fluorescent X-rays or Auger electrons will be emitted depending upon the atomic number of the daughter nuclide. The X-ray can escape from the scintillator and if the Auger electrons have very low energies then dE/dx effects are very important in calculating the counting efficiency. The number of possible interactions is very large and ignoring capture in the M and higher shells and assuming a single energy for all shells then 15 interactions are possible as shown in figure 3. However many of the interaction probabilities are small and taking ^{55}Fe as an example the 5 most important are given in table IV. In the table the coincidence counting efficiency, ϵ_i, for each interaction is

$$\epsilon_i = \phi_i (1 - e^{-n})^2$$

where ϕ_i is the interaction probability. More details are given elsewhere (16). If one assumes that all the energy is deposited at 5.89 keV ignores the dE/dx effect then $\epsilon_c = 0.898$ and allowing for the dE/dx effect $\epsilon_c = 0.734$ both of which are significantly greater than the value in the table of $\epsilon_c = 0.644$. The theoretical result will be compared with experimental data below.

TABLE IV

Calculation of the Coincidence Counting Efficiency

for ^{55}Fe with P $= 0.5\,e\,keV^{-1}$

Interaction	Electron Energy keV	dE/dx Factor F(E)	E.F(E).P n	ϵ_i'	ϕ_i	ϵ_i
(i) K X escape						
L Auger	0.64	0.378	0.121	0.013	0.013	0.000
(ii) K X capture	5.61	0.650	1.823			
L Auger	0.64	0.378	0.121			
C Auger	0.28	0.307	0.043			
			1.987	0.745	0.267	0.199
(iii) K Auger	5.24	0.641	1.679			
L Auger	0.64	0.378	0.121			
L X capture	0.37	0.330	0.061			
C Auger	0.28	0.307	0.043			
			1.904	0.724	0.001	0.001
(iv) K Auger	5.24	0.641	1.679			
L Auger	0.64	0.378	0.121			
L Auger	0.64	0.378	0.121			
			1.921	0.729	0.608	0.443
(v) L Auger	0.67	0.382	0.128	0.014	0.110	0.001
					0.999	0.644

Efficiency, $\epsilon_i = \phi_i\,\epsilon_i'$

Experimental comparison between ^{55}Fe and ^{3}H

In order to establish the theory discussed above it was decided to examine the apparent paradox that the counting efficiency for ^{55}Fe (ϵ_F) can be greater or less than that for ^{3}H (ϵ_H) for different quenching conditions (or values of P). The theoretical technique is as discussed above for ^{55}Fe and as discussed elsewhere (10) for ^{3}H. The experimental technique was as follows. Two alternative scintillator solutions were used (a) 14 cm^3 of NE220 (Nuclear Enterprises) + 0.1 cm^3 of standard solution and (b) 14 cm^3 of BBS-toluene mixture + 1 cm^3 of standard solution. The counting efficiency for ^{3}H$_2$O was 35% and 28% respectively for the two scintillators. ^{55}Fe (NO$_3$)$_3$ was dissolved in 0.1N HNO$_3$. The standard solutions were obtained from TRCL (Amersham) and have standard errors on the mean of 1.4% for ^{55}Fe and 1.3% for ^{3}H. The counting efficiency of the solutions was reduced by adding successive small aliquots of acetone.

All experimental measurements were made with a Packard Tri-Carb (Model No. 3003) at a temperature of 2°C. Integral bias curves were obtained and extrapolated to zero

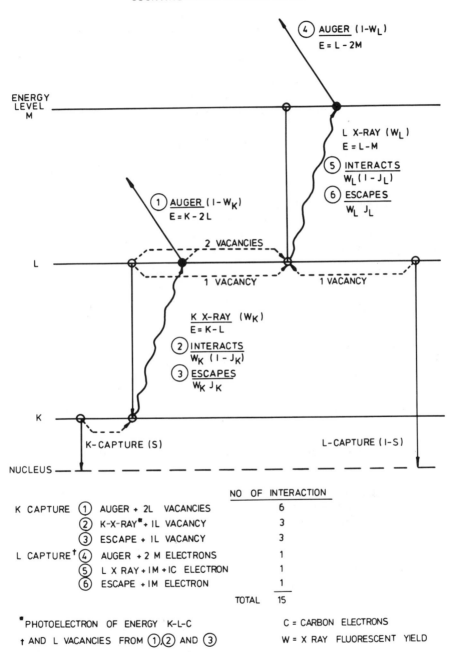

FIG.3. SCHEMATIC DIAGRAM OF ASSUMED INTERACTIONS
IN ELECTRON CAPTURE

bias. An automatic external standard was used to check that the counting efficiency for ^{55}Fe and ^{3}H was the same after the addition of the acetone aliquot and if necessary small corrections ($< 2\%$) were made.

The ratio of the counting efficiencies (ϵ_F/ϵ_H) are plotted in figure 4 as a function of ϵ_H. The ratio is for either the NE220 solution using both isotopes in separate vials or for the BBS-toluene mixture again using both isotopes. Only the calculated counting and volumetric standard deviation is shown on the figure and there is an additional systematic standard error on the mean of $\pm 2\%$ due to uncertainties in the standards. The solid line in figure 4 is that obtained from the theoretical calculations discussed above. The line has not been normalised. The agreement between the theoretical and experimental results in figure 4 is very good and supports the approach to calculating counting efficiencies proposed by the author. It is seen from the curve that the effect of quenching on the ϵ_F/ϵ_H ratio depends upon the figure of merit of the system and it is only for a single tube with $P > 0.55e$ keV^{-1} that ϵ_F/ϵ_H will increase as quenching is increased.

The reason for the change in slope of the curve is that at high counting efficiencies all the electrons produced by ^{55}Fe (figure 2) will be counted and if, for example, $n = 6$ then $\epsilon_F' = (1-e^{-6})^2 = 0.995$. With a large chemical quenching factor, (equation (5)) eg $g = 0.5$ then $n = 0.5 \times 6$ and $\epsilon_F' = 0.902$ which is only a small change (9%) in efficiency. (The counting efficiency will be reduced due to the escape of K.X-rays and the above figures for ϵ_F' should be multiplied by 0.875 to obtain the true counting efficiency, ϵ_F). However for ^{3}H, about half of the β rays have energies less than 5.9 keV and these will be more susceptible to changes by quenching resulting in a more rapid reduction in counting efficiencies ($\epsilon_H < 0.6$) than the counting efficiency for $n = 2$ is $\epsilon_F' = 0.75$ and for $g.n = 1$, $\epsilon_F' = 0.40$ (again both are to be multiplied by 0.875). In this case the β particles below 5.9 keV for ^{3}H are only a minor contributor to ϵ_H and so ϵ_H reduces more slowly than ϵ_F for increasing levels of quenching. Thus depending on the figure of merit for the system the ratio ϵ_F/ϵ_H can increase or decrease with the changes in quenching levels.

Horrocks (17) suggested that for the BBS-toluene mixture the ^{55}Fe may not be dissolved in the scintillator but it is dispersed as sols. These droplets would discriminate against the low energy Auger electrons and so reduce the ^{55}Fe counting efficiency relative to that for the higher energy β particles of ^{3}H. Experiments to measure the droplet size showed that at the concentrations used (0.6% W/W) they were less than 10 nm in diameter. Advice from sol physicists and information in the literature (18, 19) suggests that the droplet size will not be greater than 5 nm. The range of electrons in water is as given below.

Electron Energy keV	0.28	0.65	5.24	5.9
Electron Range nm	7	28	830	1000

Thus it is suggested that only the lowest energy electrons will be effected and the similarity between the separate results for the two scintillators and with the theory tends to confirm this hypothesis.

COUNTING OF OTHER RADIONUCLIDES

There are a range of EC nuclides and Horrocks (20) suggested a list which could be useful in tracer studies etc. This list is given in table V together with their calculated counting efficiencies. The reduction in counting efficiency for ^{97}Tc is due mainly to the escape of K X-rays as the fluorescent yield is 73% for Mo. Above this atomic number the fluorescent yield increases to 97% for Tl but the L X-rays (and Augers) are counted with increasing efficiency. Corresponding counting efficiencies for ^{3}H, ^{14}C and ^{36}Cl are included for comparison and efficiencies for any other isotope with a known decay scheme (and β spectrum) can be calculated by this method.

TABLE V

Counting Efficiencies for EC and β^- Nuclides
for P = 0.50 and 0.25 e keV^{-1}

Nuclide	Half-life	K_α X-ray or β max Energy keV	K capture Probability	Coincidence Counting Efficiency %	
				P = 0.5	P = 0.25
^{37}Ar (EC)	35d	2.6	0.91	23	8
^{55}Fe (EC)	2.6a	5.9	0.89	64	34
^{71}Ge (EC)	11.4d	9.2	0.88	76	55
^{97}Tc (EC)	2.6 x 10^{6}a	17.5	0.84	57	47
^{131}Cs (EC)	9.7d	29.8	0.87	70	52
^{179}Ta (EC)	600d	55.8	0.63	87	53
^{205}Pb (EC)	3 x 10^{7}a	12.2	*	95	72
^{3}H (β)	12.26a	18.6	–	57	35
^{14}C (β)	5730a	156.	–	95	91
^{36}Cl$^+$ (β)	3.1 x 10^{5}a	714.	–	99	98

*Only 35 keV available for the decay.

+1.9% EC $K_{\alpha 1}$ = 2.3 keV.

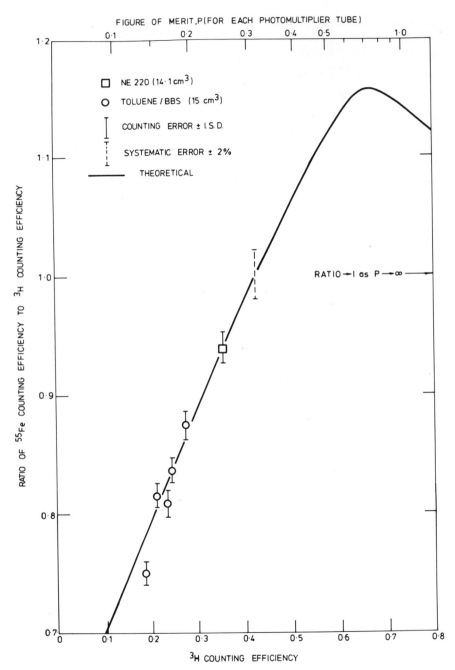

FIG.4.RATIO OF $^{55}Fe/^{3}H$ COUNTING EFFICIENCIES FOR A
COINCIDENCE SYSTEM

Determination of plutonium alpha activity and ^{241}Pu in biological materials

This is a particular problem of the nuclear energy industry where levels of soluble plutonium in man can only be determined by urine and faecal counting. The additional problem is the increasing levels of ^{241}Pu due to increased burn-up of fuel in reactors. ^{241}Pu is a β-emitter (E max = 20.8 keV, $T_{1/2}$ = 13.2a). After extraction of the plutonium from the biological material then liquid scintillation counting offers a convenient method of determining both the α activity and the ^{241}Pu β activity simultaneously.

Horrocks and Studier (21) first described the technique for ^{241}Pu in urine. Eakins and Lally (22) have a relatively simple technique of gel scintillation counting for Pu α and β^- activity determination. The α activity from ^{239}Pu (+ ^{238}Pu and ^{240}Pu) can be counted with 100% efficiency but it is convenient to reduce it to 90% to count the ^{241}Pu at the same time with a counting efficiency of 21%. This technique is used to detect 1 pCi of α activity and 10 pCi of ^{241}Pu at the background level. The limits of detection are about 20% of these levels. The technique is adequate for the ^{241}Pu counting in urine, faeces and nose blow samples but to obtain adequate sensitivity for ^{239}Pu in urine (0.02 pCi) it is still necessary to measure electrodeposited sources in a solid-state counter.

CONCLUSION

Liquid scintillation counting is the accepted technique for many radio isotopes in a wide range of chemical forms. It lacks sensitivity for very low levels of α-activity and cannot compete directly with the internal gas counter for natural levels of tritium. The technique has come a long way since the first major conference recorded by Bell and Hayes in 1958 (7). It is as important now as it was 20 years ago to establish an understanding of the underlying theory behind the techniques so that improvements can be made and pitfalls avoided. Electron capture nuclides are valuable tools in all types of tracer techniques but they can produce unusual results unless a proper understanding is obtained.

REFERENCES

1.	D.L. Horrocks, Applications of Lqiuid Scintillation Counting. New York and London: Academic Press (1974).

2.	A. Dyer, Ed, Liquid Scintillation Counting, Vol 1. London, New York and Rheine: Heyden & Son Ltd. (1971).

3.	M. Crook, P. Johnson and B. Scales, Ed, Liquid Scintillation Counting, Vol 2. London, New York and Rheine: Heyden & Son Ltd. (1972).

4.	M.A. Crook and P. Johnson, Ed, Liquid Scintillation Counting, Vol 3. London, New York and Rheine: Heyden & Son Ltd. (1974).

5.	E.D. Bransome, Ed, The Current Status of Liquid Scintillation Counting, New York and London: Grune and Stratton (1970).

6. D.L. Horrocks and C-Z. Peng, Ed, Organic Scintillators and Liquid Scintillators. New York and London: Academic Press (1971).

7. C.G. Bell and F.N. Hayes, Ed, Liquid Scintillator Counting. London, New York, Paris and Los Angeles: Pergamon Press (1958).

8. J.A.B. Gibson and A.E. Lally, Analyst 96, 681 (1971).

9. H.J. Gale and J.A.B. Gibson, J. Sci. Instrum. 43, 224, (1966) and AERE Report R 507 (1965).

10. J.A.B. Gibson and H.J. Gale, J. Sci. Instrum. (J. Phys. E) Ser 2, 1, 99 (1968).

11. J.A.B. Gibson and H.J. Gale, Int. J. Appl. Rad. & Isotopes, 18, 681, (1967).

12. J.A.B. Gibson in Liquid Scintillation Counting, Vol 2, p 23. (M.A. Crook, P. Johnson and B. Scales, Ed). London, New York and Rheine: Heyden and Sons (1972).

13. J.A.B. Gibson. AERE Report R 6919 (1972).

14. J.B. Birks. The Theory and Practice of Scintillation Counting, Oxford. Pergamon Press (1964).

15. F.E.L. ten Haaf in Liquid Scintillation Counting, Vol 2, p 39, (M.A. Crook, P. Johnson and B. Scales, Ed). London, New York and Rheine: Heyden & Son (1972).

16. J.A.B. Gibson and M. Marshall. Int. J. Appl. Radiat. & Isotopes, 23, 321, (1972).

17. D.L. Horrocks, Private communication.

18. S. Glasstone, Text book on Physical Chemistry p 1267. London: MacMillan, (1948).

19. N.K. Adam, Physical Chemistry, p 609. Oxford: Clarendon Press (1958).

20. D.L. Horrocks. Int. J. Appl. Radiat. & Isotopes, 22, 258, (1971).

21. D.L. Horrocks and M.H. Studier, Anal. Chem. 30, 1747, (1958).

22. J.D. Eakins and A.E. Lally in Liquid Scintillation Counting, Vol 2, p 155 (M.A. Crook, P. Johnson and B. Scales, Ed). London, New York and Rheine: Heyden & Son (1972).

CERENKOV COUNTING AND LIQUID SCINTILLATION COUNTING

FOR THE DETERMINATION OF ^{18}FLUORINE

D.N. Abrams, S.A. McQuarrie, C. Ediss, and L.I. Wiebe
Division of Bionucleonics and Radiopharmacy
University of Alberta
Edmonton, Alberta

Abstract

The positron emitted (E_{max} 635 KeV) upon radioactive decay of ^{18}F has been used to measure ^{18}F in liquid scintillator solutions which have low or high water capacity, and in hydrophilic and hydrophobic liquids which have different refractive indicies. The influence of nitromethane, a chemical quenching agent, on counting efficiency in each of these liquids has been measured, and is discussed on the basis of observed shifts in the pulse height spectra. The contribution of coincident Compton events arising from the annihilation gamma rays, to the overall counting efficiency, has been estimated using methyl salicylate as the counting medium.

Introduction

Fluorine-18 is a short lived ($T_{\frac{1}{2}}$ = 110 min) radionuclide which decays primarily (97%) by the emission of positrons (β^+; E_{max} 635 KeV), and some (3%) electron capture.[1] Annihilation γ rays (511 KeV, 194%) and x-rays are associated with these respective decay modes. It is the coincident γ-rays that are most frequently used in the radioassay of this and other positron emitting radionuclides.

Although most radioisotopes analysed by liquid scintillation counting are either α or β^- emitters[2], and those assayed by Cerenkov counting in liquid solution are β^- emitters[3], positrons can be counted equally well by these methods.

We now report preliminary observations from the radioassay of ^{18}F using liquid scintillation fluors (toluene/PPO/POPOP, and Aquasol$^{T.M.}$), water and methyl salicylate as photon generating media for counting with commercial liquid scintillation spectrometers.

Experimental
    Fluorine-18 was produced by irradiation of ^{20}Ne gas
with 6.5 Mev deutrons using the 7 MV Van de Graaff electrostatic generator at the University of Alberta. Fluorine-18
was recovered from the glass target vessel with a methanol
flush. Carrier NaF (1 mg ml^{-1}) was added to that solution
prior to liquid scintillation radioassay. The ^{18}F content
of an aliquot of the final solution was standardized by
γ-γ coincidence counting of the 511 Kev annihilation photons,
using a 1¾"x2" NaI well crystal detector. After cross
calibration with ^{137}Cs (11% efficiency), a final counting
efficiency of 35 ± 5 percent was calculated. This value
was used to estimate the counting efficiency of the liquid
scintillation (LS) and Cerenkov (C) procedures.
    Aliquots (10 µl) of the Na^{18}F-methanol solution were
accurately pipetted into glass scintillation vials (Kimble,
Toledo) containing 15 ml of either distilled water, methyl
salicylate (MS) (redistilled, Reagent Grade, Fisher
Scientific, Edmonton), Aquasol$^{T.M.}$ or toluene containing
PPO (4 g l^{-1}) and POPOP (50 mg l^{-1}) (Scintillation grades,
Fisher Scientific, Edmonton). Samples were counted in the
LS spectrometers, and then all but the water samples were
quenched with 200 µl of nitromethane (reagent grade, Fisher
Scientific, Edmonton) and counted again. Water (200 µl)
was added to one Aquasol sample as an alternate quenching
agent.
    To determine the contribution of pulses arising as a
result of the interaction of the 511 Kev coincident annihilation γ's, a liquid scintillation vial was fitted with
a stainless steel cylindrical tube which was open at the
upper end but sealed at the bottom (Figure 1). An aliquot
(10 µl) of the Na^{18}F/methanol solution was introduced into
the tube, and the tube was then inserted in a LS vial
containing 15 ml of MS. The sample was counted, the tube
removed, and then recounted to ensure that the generating
liquid was free of ^{18}F contamination. That vial was then
recounted containing 10 µl of the Na^{18}F methanol intimately
mixed with the MS in the vial.
    All samples were counted using both the Liquimat 220
(Picker Nuclear, New York) and the Mark III (Searle Instrumentation, Chicago) LS spectrometers. Pulse height spectra
(PHS) were stored in a Northern Scientific NS636 multichannel analyzer.
    All count rates were decay corrected to the time of
standardization of the Na^{18}F solution.

Table 1. Counting efficiency and figure of merit (E^2/B for ^{18}F in toluene/POP/POPOP (TPP), Aquasol$^{T.M.}$, methyl salicylate (MS) and water, using the Picker Liquimat 220 and the Searle Mark III LS spectrometers. NM indicates the addition of 200 μl of nitromethane, H_2O the addition of 200 μl of water. Counting precision is estimated to be ± 10 percent.

Generating System	Liquimat 220 % E	Mark III (^{32}P window) % E	E^2/B
TPP	80.7	78.1	135.6
TPP + NM	53.1	56.7	
Aquasol	98.2	99.3	200.8
Aquasol + H_2O	98.7	98.5	
Aquasol + NM	87.0	90.3	
MS	78.8	85.9	282.7
MS + NM	70.2	76.6	
H_2O	2.3	3.7	0.41

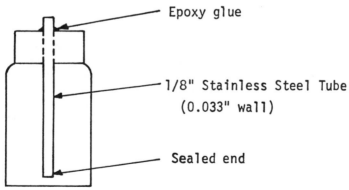

Epoxy glue

1/8" Stainless Steel Tube (0.033" wall)

Sealed end

Figure 1. Modified liquid scintillation vial for determination of Compton electron contribution to ^{18}F Cerenkov pulse height spectrum.

Results and Discussion

Counting efficiencies observed for ^{18}F in the various media both with and without the addition of the chemical quenching agent (nitromethane) are presented in Table 1, and Pulse height spectra for ^{18}F in TPP, Aquasol and MS are depicted in Figure 2. Of particular interest are the high counting efficiencies (essentially 100%) observed using Aquasol and the relatively high counting efficiencies (78.8%) using MS, compared with TPP (80.7%). MS has been shown to be an excellent Cerenkov counting medium, with its high refractive index (1.522) and wave-shifting character-istics.[4] Counting efficiencies for [36]Cl (β^- E$_{max}$ 714 Kev) in MS were 82.2 and 91.6 percent of TPP values respectively for the Liquimat 220 and Mark III LS spectrometers. Further-more, Aquasol has been found to be less efficient than TPP when counting weak β^- emitters[5], and TPP has been widely used to obtain efficiencies approaching 100 percent for most energetic β^- emitting nuclides. Dissolution of Na[18]F is a possible explanation of the phenomenon. The low MS background, even in the wide window necessary for LS counting, gives the MS system a distinct advantage over the LS systems in terms of figures of merit (E^2/B) (Table 1).

The annihilation gamma photons were found to produce a count rate of 9.6 percent of that observed when the [18]F (hence the β^+s) was dissolved directly in the methyl sali-cylate. Although pulses derived via the gamma interactions would not likely contribute to the count rate, they could well decrease the sensitivity of the system to chemical or to color quench.

The counting efficiency for ^{18}F in water was found to be too low to be of any practical interest. The PHS observed was similar to that reported for ^{32}P in water, using the same LS spectrometer.[6]

Figures 3 and 4 depict the pulse height spectra obtained for ^{18}F positrons and ^{137}Cs compton electrons (External standard) in MS, using the Liquimat 220 LS spectrometer. It is apparent that the pulse height ranges are virtually identical, although there are qualitative differences in the nature of the spectral shifts when the systems are quenched. It has been shown that a sufficiently high count rate can be obtained from the [137]Cs Liquimat 220 external standard source to allow precise standardization for [36]Cl samples by the ESCR method.[4] Similarities between the ^{18}F and [36]Cl PHS in MS would lead to the conclusion that ESCR quench correction would be ideal for ^{18}F in MS.

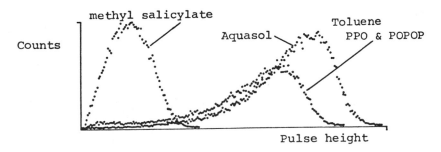

Figure 2. Pulse height spectrum of ^{18}F positrons in methyl salicylate, aquasol & toluene PPO & POPOP.

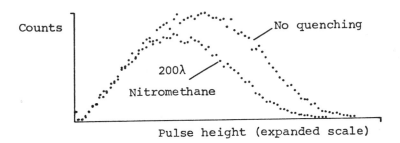

Figure 3. Pulse height spectrum of ^{18}F positrons in methyl salicylate with and without quenching.

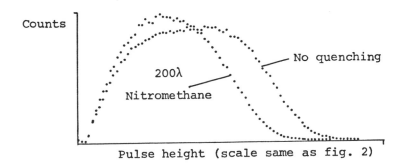

Figure 4. Pulse height spectrum of the Liquimat 220 ^{137}Cs external standard in methyl salicylate with and without quenching.

References
1.   C.M. Lederer, J.M. Hollander, and I. Perlman, Table of
      Isotopes, 6th Ed., J. Wiley and Sons, N.Y. (1968) p. 7.
2.   A.A. Moghissi, in The Current Status of Liquid Scintil-
      lation Counting, E.D. Bransome, Jr. (Ed.), Grune and
      Stratton, N.Y. (1970) p. 86.
3.   H.H. Ross and G.T. Rasmussen, in Liquid Scintillation
      Counting: Recent Developments, P.E. Stanley and B.A.
      Scoggins (Eds.), Academic Press, N.Y. (1974) p. 363.
4.   L.I. Wiebe and C. Ediss, Int. J. Appl. Radiat. Iso-
      topes (in press) (abstract).
5.   L.I. Wiebe, A. Stevens, A.A. Noujaim and C. Ediss,
      Int. J. Appl. Radiat. Isotopes, 22, 663 (1971).
6.   L.I. Wiebe, A.A. Noujaim and C. Ediss, Int. J. Appl.
      Radiat. Isotopes, 22, 463 (1971).

AN INTERFACE FOR ROUTINE SPECTRAL DISPLAY FROM SEVERAL LIQUID SCINTILLATION COUNTERS

B. E. Gordon[a], M. Press, W. Erwin, and R. M. Lemmon
Laboratory of Chemical Biodynamics
Lawrence Berkeley Laboratory
University of California, Berkeley, CA 94720

Abstract

An interfacing circuit for coupling several Tri-Carb liquid scintillation counters to a multichannel analyzer is described. The display of the analyzer is completely controlled from the liquid scintillation counters and information from any one of the three channels in each counter can be displayed.

The system is being used for both instructional and diagnostic purposes. For the former it will be part of a course on the theory and use of liquid scintillation counting. For the latter it assists in setting up optimum counting conditions for single and double labels in new cocktails as well as for new labels, and to help diagnose counting problems such as chemi- and/or photoluminescence.

Introduction. Work at our laboratory involves about 100 people, many of whom employ liquid scintillation counting. Because this is both an instructional and a research institution, there is a high annual turnover rate, so that newcomers employing radioisotopes must learn to use liquid scintillation counting properly before embarking on their projects. For this reason a short course is taught annually covering the theory and use of liquid scintillation counting as well as sample preparation and data handling methods. Visual display of spectra obtained under a variety of realistic counting conditions would be a valuable adjunct to the course; this was the primary motive for the work reported here.

In addition, we have long had a need for rapidly, and simply, generating beta spectra of samples from new compounds that contain any of a variety of isotopes (*e.g.*, ^{14}C, ^{3}H, ^{3}P, ^{33}P, ^{36}Cl, ^{51}Cr, ^{125}I) and that require specific liquid scintillation cocktails. This would permit us to more intelligently set the optimum counting conditions than is now the case, and thus would be a significant improvement over the present method of blindly probing for the best arrangement of gain, window location, and width.

Fairman and Sedlet (1) briefly described a multichannel analyzer (MCA) coupled to a liquid scintillation counter (LSC)
a) author to whom correspondence should be addressed.

with a logarithmic energy scale (1). This was used to help
develop a method for the analysis of a ^{210}Pb, ^{210}Bi, ^{210}Po
mixture. No details of the interfacing circuit were supplied.
Neary and Budd (2) employed a similar system capable of operating in the logarithmic or linear output mode. They studied
the differences in beta spectra arising from chemical and
color quenching. No details on interfacing of the MCA and LSC
were given. Klein and Eisler (3) also coupled an MCA with an
LSC to determine the effect of color vs. chemical quenching
and used the cathode ray tube to visually project the altered
spectra. They used the MCA amplifier to obtain a linear display of the energy scale and eliminated the upper discriminator in order to display the full spectra.

Experimental.   The above reports employed Beckman counters.
Our laboratory has three Packard Inst. Co. counters, two
Model 3375 and one 3385. We understand that liquid scintillation counters from this company are in use in substantial
numbers around the world. Thus it seemed worthwhile to develop an interfacing system for this make. We proceeded along
the same line as described in reference (1), *i.e.*, the signal
to be presented to the multichannel analyzer was picked off
after the summation circuit and linear amplifier but immediately before the upper and lower discriminators. The gating
pulse, *i.e.*, the one used to determine which signal pulse was
to proceed to the multichannel analyzer, was obtained from
the ratemeter output jack located on the rear of the counter.
Thus, the requirement that the signal and gating pulses be in
coincidence meant that the pulse analyzed (in the MCA) was
the pulse counted (in the LSC). Furthermore, this arrangement
meant that the effect of altering the gain or discriminator
settings on the LSC would appear in the spectral display.

Unfortunately, presentation of the unaltered pulses to the
coincidence circuit of several commercial multichannel analyzers tested (3 in all) produced no useful spectra. The problem was that the pulses put out by the liquid scintillation
counters did not match the coincidence input pulse requirements of the MCA's. We chose, finally, the Northern Scientific Model NS-700 largely because this MCA is in use by other
groups in the laboratory.

The differences in output signals by the LSC's and input
requirements of the MCA are: the former puts out negative
pulses (from the summing network) having rise, width, and
fall times in the 30-100 nsec range while the MCA requires
positive going pulses of 100 nsec rise and fall and 1000 nsec
width.

174

Alteration of the input amplifier of the MCA could have
been performed, but this would have interfered with other
applications of this analyzer. Therefore, an additional
amplifier board was purchased from the Northern Scientific
Co. and modified so as to both invert and stretch the LSC
pulses to match the input requirement of the coincidence section of the MCA. Similarly, the gating (ratemeter) pulse was
stretched so that its shape was very close to the shaped signal pulse. A simple block diagram of the system is shown in
Fig. 1. The essential features of the interface are:

1) *Signal routing*. The signal is taken from each pulse
height analyzer module at test point (see Packard schematic
D7100003 linear amplifier emitter-follower output) and fed
through coaxial cable to an attenuator of approximately 20:1.
The attenuator output is matched to a 50 ohm coaxial line
which is routed to the interface box where one of nine similar cables is selected by a rotary switch (channel selector).
The signal is then fed to the input of the interface ampli-
fier/shaper through a 100 ns risetime integrator terminated
in 50 ohms. The integrator smooths any cable ringing or signal ringing originating in the LSC.

2) *Interface amplifier/shaper*. A standard Northern Scientific preamp/amplifier board (NS-700/S-129) was slightly
modified for use as the interface amplifier/shaper. Voltage
gain of the first stage was reduced from approximately 3300
with a 10 µs time constant to approximately 7.5 with no capacitor in the feedback loop. Output from this stage is fed to
a 2K ohm front panel mounted "gain control" potentiometer.
The second stage is fed from the slider of the gain control
and is unmodified with a non-inverting gain of 2. The third
stage has been modified for two voltage gain settings, front
panel selectable, of 0.9 or X9 inverting. Normally the X9
gain setting is used unless very high level input signals are
encountered (e.g., ^{32}P). The last stage remains unmodified
with a voltage gain of approximately 5 inverting and a 47 pf
integrating capacitor across the feedback resistor. The
shaped and amplified signal is then applied to the signal out
connector through a 33 µfd capacitor.

Gating pulse stretcher. Positive spikes from the ratemeter output jacks (9 in all) are fed through coaxial lines
to the second wafer of the channel selector switch. A spike
from the appropriate pulse height analyzer is then inverted
and used to trigger a variable width multivibrator oneshot
whose output is buffered and fed to the coincidence output
jack. The oneshot width is screwdriver adjusted via a front
panel mounted trimpot to match the width of the positive side
of the signal from the signal-out jack. Pulses to the NS-700

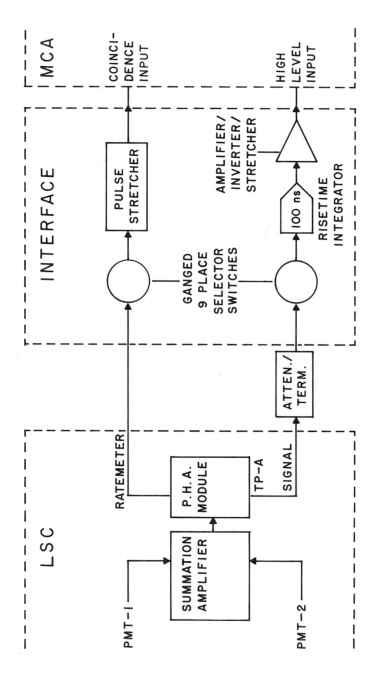

Figure 1. Block Diagram of LSC/MCA Interface.

coincidence input are now of sufficient width and level to operate internal gating circuitry.

Recalling that the system had to be able to interrogate nine channels from three LSC's, it is evident that considerable cabling was employed. Problems associated with the cables (ringing, attenuation, distortion) were solved by 50 ohm termination. Nevertheless, the original signal from the LSC's was rather distorted (Fig. 2a) and resulted in a sharply distorted tritium spectrum (Fig. 2b). The introduction of a 100 nsec rise time integrator on the input circuit smoothed the pulses very well (Fig. 2c) and this in turn resulted in an acceptable tritium spectrum (Fig. 2d).

To be sure that the MCA was viewing exactly the same portion of the spectrum as the LSC, the added requirement was made that the count rates in both instruments agree. It was our experience that normal-appearing spectra were sometimes obtained where the count rate in the MCA differed significantly from that in the LSC--generally, but not invariably, lower. The gain control on the interface circuit served to achieve this end by increasing gain to avoid weak signal loss down the cable or decreasing gain to avoid overload of the interface amplifier with resultant clipping. The interface box is shown in Fig. 3, with the fast rise time integrator on the right and the pulse shaper in the center. Fig. 4 shows the assembled system with spectra from a ^{3}H channel and ^{14}C channel.

Results and Discussion. With this system in hand, some studies were made of the spectral shapes of samples counted under various conditions. Figure 5 shows the spectrum of ^{3}H + ^{14}C in the tritium and carbon-14 channels. Note the tritium "spike" on the low energy part of the carbon-14 spectrum in Fig. 5a. In each channel, one of the LSC discriminators can be visually adjusted to give a satisfactory setting for one isotope in the presence of the other. For the best separation one may still have to adjust each discriminator (the upper level in the ^{3}H channel and the lower level in the ^{14}C channel) stepwise, getting the efficiency and "spillover" each time. The visual setting permits one to narrow the range of interest, a feature of great benefit to the inexperienced user.

One of the most annoying problems faced in biological research with radiotracers is that of chemiluminescence. It is normally detected by making repeated counts to see if the rate decreases with time. Because of its low energy, one should observe this readily in the tritium channel as a low energy spike. Chemiluminescence was generated by adding a few

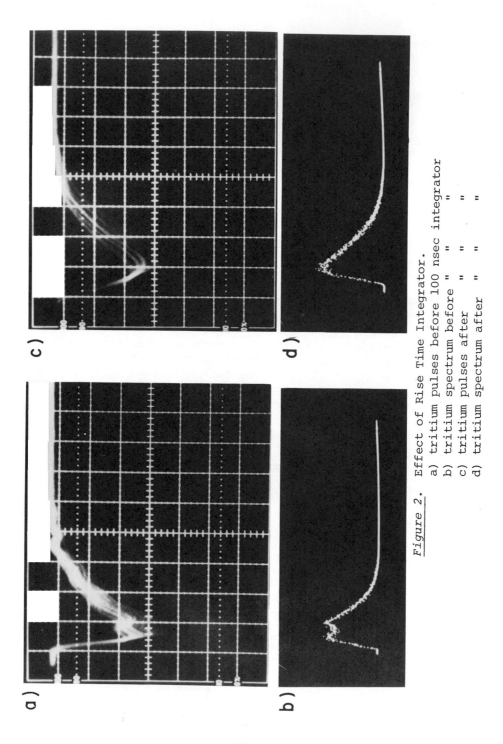

Figure 2. Effect of Rise Time Integrator.
a) tritium pulses before 100 nsec integrator
b) tritium spectrum before "   "   "   "
c) tritium pulses after   "   "   "   "
d) tritium spectrum after   "   "   "   "

178

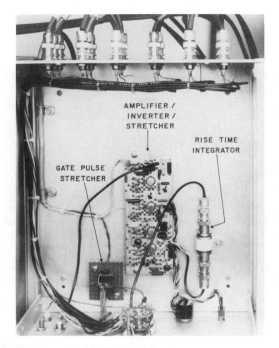

Figure 3. Interface with Top Removed.

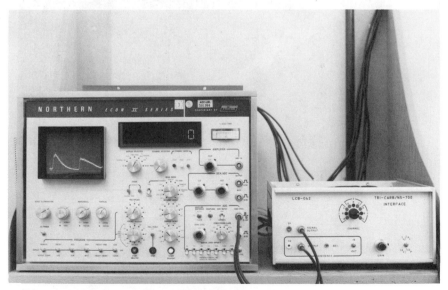

Figure 4. Multichannel Analyzer Plus Interface.
Tritium (left) and carbon-14 (right) spectra
displayed.

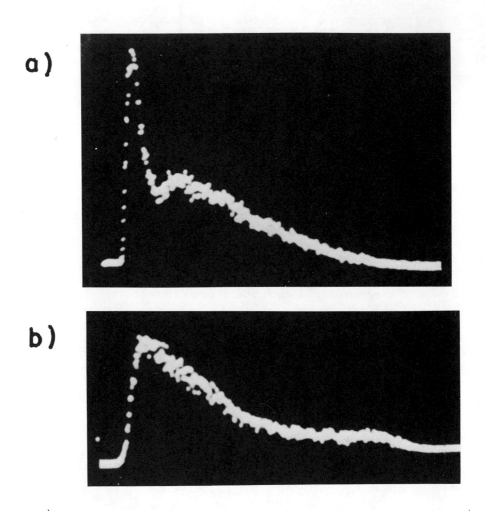

Figure 5. Doubly Labeled Sample.
a) $^3H + ^{14}C$ in ^{14}C channel
b) $^3H + ^{14}C$ in ^3H channel

drops of a strong base to a commercial detergent cocktail and the spectrum shown in Fig. 6a was produced. This is compared with a lightly quenched ^3H sample (40% efficiency). The difference is readily apparent. Because chemical quenching shifts the beta spectrum to a lower energy, one would expect that for highly quenched samples, the presence of chemiluminescence would be difficult to distinguish from tritium. This is indeed the case, as shown in Fig. 6b. There is no obvious difference in the shape of the ^3H spectrum at 10% counting efficiency from the chemiluminescent spectrum.

The MCA can also operate in a multi-scaling mode, *i.e.*, where a count is made for a preset time, stored in channel 1, repeated, stored in 2, etc. Thus, count rate vs. time can be visually demonstrated, if one wishes, to show either chemi- or photoluminescence, as is demonstrated in Fig. 7. It is readily apparent that the half lives of the two processes are entirely different and that one can return to background in a few minutes with photoluminescence, but not for several hours with chemiluminescence.

The decay of chemi- or photoluminescence to true background is much more important for low count rate (*i.e.*, ∿100 cpm) samples because both of these processes have complex kinetics and there may be a very long-lived tail (particularly for photoluminescence) in the decay curve, which will affect the results. Such low count rates are not readily studied by the above system because they require an inordinately long counting period to collect enough counts in each channel to generate a useful spectrum or decay plot. This, of course, is a basic limitation of the MCA-LSC system, so it is normally not used for low count rate samples.

When setting up the system to study a spectrum, we have found that the greater the isotope energy, the lower the interface gain should be. The proper adjustment depends on each isotope, but the interface gain must be set so that the count rate of the LSC and MCA agree. A rule of thumb has evolved from use of this system: the interface gain adjusts the beta spectrum so that the high energy end is between 25 and 50% of full scale. Less than this results in a lower count rate in the MCA due to low energy pulse loss; much greater than this yields a distorted spectrum due to high energy pulse clipping. This may not be an inviolate rule for all isotopes, but was found to be so for tritium and carbon-14.

a)

b)

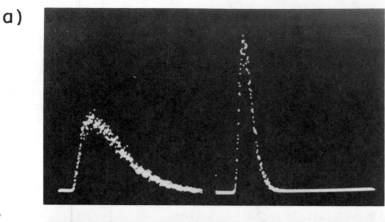

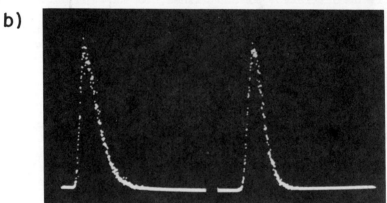

Figure 6.   Tritium Spectra vs. Chemiluminescence
a) ^{3}H (left), chemiluminescence (right) –
40% ^{3}H efficiency
b) ^{3}H (left), chemiluminescence (right) –
10% ^{3}H efficiency

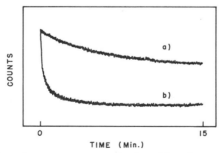

Figure 7.   Decay Rates of Chemiluminescence (a) vs.
Photoluminescence (b).

Conclusion. An interfacing circuit has been developed for a routine multi-channel analysis of radioisotope spectra from a number of liquid scintillation counters. A selector switch permits the selection of any one of nine channels from three liquid scintillation counters. The consequences of changing gain and/or discriminator settings are immediately reflected in the displayed spectrum.

The system permits one to observe the presence of chemi- or photoluminescence on new samples. It also is helpful in finding the appropriate liquid scintillation counter settings for doubly labeled samples. It is particularly useful for helping to set up counting conditions for radioisotopes not previously used. It promises to be a valuable instructional aid to those encountering liquid scintillation counting for the first time.

Schematics of the interface and associated cabling are available on request.

Acknowledgements. The prototype interface circuit was constructed by R. Healey of Healey Associates, Dublin, Calif. and much of the construction of the final system was done by John Griffin of this laboratory. The work reported in this paper was supported by the U.S. Energy Research and Development Administration.

References

1. W.W. Fairman and J. Sedlet, Anal. Chem. <u>40</u>, 2004 (1968).

2. M.P. Neary and A.L. Budd <u>in</u> The Current Status of Liquid Scintillation Counting, Chapt. 28 (Edwin D. Bransome, Ed.) New York and London: Grune and Stratton (1970).

3. P.D. Klein and W.J. Eisler <u>in</u> Organic Scintillators and Liquid Scintillation Counting, pp. 395-418 (D.L. Horrocks and C.T. Peng, Eds.). New York: Academic Press (1971).

ABSOLUTE DISINTEGRATION RATE DETERMINATION OF BETA-EMITTING RADIONUCLIDES BY THE PULSE HEIGHT SHIFT-EXTRAPOLATION METHOD

by
Donald L. Horrocks
Scientific Instruments Division
Beckman Instruments, Inc.
Irvine, California 92713

It has long been the hope of investigators using liquid
scintillation counters to be able to determine the disintegration rate of a radionuclide in a sample directly, without
having to use a standard solution of the radionuclide to
calibrate the counting system. In this paper, a method will
be described which enables the investigator to determine the
disintegration rate of a beta emitting radionuclide in a
sample by measuring the pulse height shift produced by quenching (either real or simulated). This paper also describes
two new liquid scintillation counters, Beckman LS-8000 and
Beckman LS-9000, which provide the capability for determination of disintegration rates by this method.

Three previous reports[1-3] have dealt with the application
of a double extrapolation method based upon the measurement
of the amount of successive increases in quench by the Compton
edge-half-height method. The first two reports dealt with
quench produced by the addition of a quenching agent directly
into the sample-liquid scintillation solution. The last
report dealt with a method of simulated quench by introduction
of an optical filter between the sample and the detector
(multiplier phototube). All three of the reports utilized a
single multiplier phototube (MPT) detection system; i.e., a
non-coincidence system. Because of the high background which
results from use of the single MPT, it was necessary to count
the sample at various threshold levels and extrapolate to a
zero threshold in order to obtain the sample count rate at
each quench level. The logarithm of the extrapolated count
rate is then plotted as a function of the relative pulse
height for the half-height of the Compton edge. Extrapolation of these plots to give a zero relative pulse height
ratio would provide the sample disintegration rate. The
disintegration rates were determined by this method for
samples of the following radionuclides: reference 1 - ^{14}C;
reference 2 - ^{14}C and ^{63}Ni; reference 3 - ^{3}H, ^{106}Ru, ^{14}C,

^{95}Nb, and ^{60}Co. The accuracy of the method was $\pm$ 2%.

NEW METHOD

This method differs from the previous in that it is accomplished in a coincidence type liquid scintillation counter and does not require the extrapolation of measured count rate to zero threshold at each quench level. The method also requires an instrument with pulse summation which gives a truer representation of the pulse height spectrum. The coincidence technique allows for the threshold to be set at essentially the threshold of detection without including excessive backgrounds. The count rates measured, with the threshold set at zero, will be the same as that obtained by the count rate extrapolation method used in previous methods.

The other requirement for this method is the measurement of a parameter which is indicative of the scintillation efficiency. In previous methods the pulse height equivalent to the half-height of the Compton edge for a gamma ray source was used to monitor the scintillation efficiency of the solution as a function of the quench level (real or simulated). In the present method, a new parameter is measured which is a direct measurement of the scintillation efficiency. This new parameter is called the H-number (H#).[4] The H# measures the pulse height of the inflection point $\dfrac{d^2 \ (CPM)}{d \ (PH)^2} = 0$ of the Compton edge. This inflection point is unique, i.e., there is only one point on the Compton edge for which the second derivative is zero. The use of the H# technique requires the use of a gamma ray source which will produce a Compton distribution that is the result of a single gamma ray energy. One ideal source for producing this type of distribution is ^{137}Cs - ^{137m}Ba which produces the 662 keV gamma rays with no other interfering gamma rays.

A plot of the ratios of the relative scintillation efficiency for a ^{3}H sample with successive additions of a quenching agent vs. the logarithm of the zero threshold count rate is shown in Figure 1. Table 1 lists the actual data obtained from the liquid scintillation counter. A least squares fit of the data to the equation:

$$\log CPM = c + d \ R$$

gives

$$\log CPM = 5.22766 - 0.23595 \ R$$

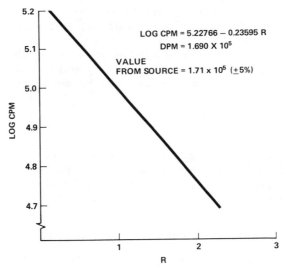

Figure 1. Extrapolation of plot of logarithm of measured CPM in wide open counting channel (zero threshold) vs. relative scintillation yield ratio (R) to obtain DPM of sample. Successive additions of quenching agent produced increased values of R.

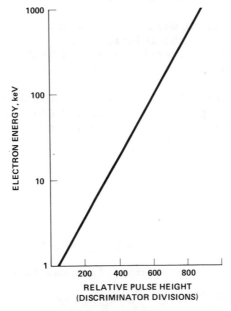

Figure 2. Pulse Height - Energy relationship for the Beckman LS-8000 and LS-9000 Liquid Scintillation Counters.

where c = 5.22766 is the logarithm of the DPM. Thus the DPM of the sample is 1.69×10^5. The actual DPM value calculated from the stock calibration is 1.71×10^5 ($\pm$ 5%). The value of c is the logarithm of the sample count rate when R is zero. When R is zero, every beta decay will produce a measurable count; i.e., the count rate will be equal to the disintegration rate.

USE OF H-NUMBER

The pulse height-energy relationship has been previously investigated.[5,6] The Beckman LS-8000 and Beckman LS-9000 liquid scintillation counters utilize a logarithmic pulse height conversion leading to a pulse height-energy relationship:

$$PH \text{ (pulse height)} = a + b \log E$$

where E is the energy of an electron producing the measured pulse height. For these systems, the pulse height response is shown in Figure 2. If E is expressed in keV, the value of a is the pulse height corresponding to a one keV electron. The equation for the response of this particular instrument used in this investigation (Note 1) was:

$$PH = 121 + 250 \log E.$$

NOTE 1. Due to use of a prototype LS-8000 for this work, the relationship for commercially available instruments may be different.

The H# is a measure of the difference in pulse height units of the inflection point of the Compton edge of any sample (PH_q) relating to the inflection point of the Compton edge of an unquenched sample (PH_o):

$$H\# = PH_o - PH_q$$

The response of the two samples is:

$$PH_o = a + b \log E_o$$

$$PH_q = a + b \log E_q = PH_o - H\#$$

Subtracting the second equation from the first gives:

$$H\# = b \log (E_o/E_q)$$

TABLE 1

SAMPLE	ZERO THRESHOLD CPM	RELATIVE SCINTILLATION EFF.	R[a] VALUE
1 [b]	98,111	1.000	1.000
2 [c]	86,582	0.808	1.237
3	75,341	0.673	1.486
4	60,666	0.534	1.871
5	50,626	0.449	2.229

[a] $R = \dfrac{\text{Relative Scintillation Eff. of Sample 1}}{\text{Relative Scintillation Eff. of measured Sample}}$

[b] Unquenched sample

[c] Samples 2-5 contain increasing amounts of quench.

where R is defined as:

$$R = E_o/E_q$$

Thus knowledge of b (an instrument parameter) and a measure
of H# is all that is necessary to calculate the value of R:

$$R = \text{antilog } (H\#/b)$$

Table 2 lists the values of the DPM of a sample obtained by
this method utilizing different values of the slope b. The
real DPM value of this sample was 97,370. The value of b is
very critical to the accuracy of the method. A known DPM
sample can be used to check on the accuracy of the value of
b.

Figure 3 shows the application of this method for a series of
samples with the same amount of ^{3}H but different amounts of
quench. The value of R for the least quenched sample is not
1.00 because it was not a totally quench free sample.
However, this method does not require that the least quenched
sample be totally quench free.

SIMULATED QUENCH

This method works equally well when the quench is artifically
created by introduction of some optical absorber between the
sample and the detectors (MPTs). Flynn, et al.[3] first
demonstrated use of simulated quench by use of calibrated
filters. However, in the present method any filter material
can be used because it is calibrated at the time the zero
threshold CPM is measured. This method has great desirability
because the sample-liquid scintillator solution remains
unaltered. Figure 4 shows the determination of DPM of a ^{3}H
sample using the simulated quench monitored by the H#. The
vial was merely wrapped with paper of different color. The
paper was typewriter paper of white, pink, blue, and yellow
color. The paper was wrapped carefully around the vial and
taped in place so as not to come lose or jam the liquid
scintillation counter elevator mechanism.

Figure 5 and 6 show the determination of the DPM of ^{14}C
samples by this method using the simulated optical and added
chemical quench. Figures 7 and 8 show similar plots for
determination of the DPM of ^{3}H-water samples in an emulsion
liquid scintillation solution using the simulated optical and
added chemical quench techniques.

TABLE 2

THE INFLUENCE OF THE VALUE OF THE SLOPE b ON THE
DETERMINATION OF THE DPM OF A ^{3}H SAMPLE. ACTUAL DPM = 97,370

SLOPE b	DPM
267	102,605
258	99,867
250	97,481
240	94,410

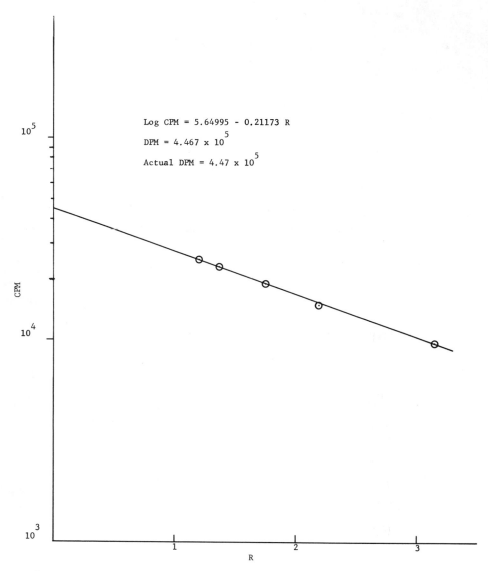

Figure 3. Extrapolation method plot applied to a set of samples with the same amount of ³H but different amounts of quench. Each R value corresponds to a different counting sample.

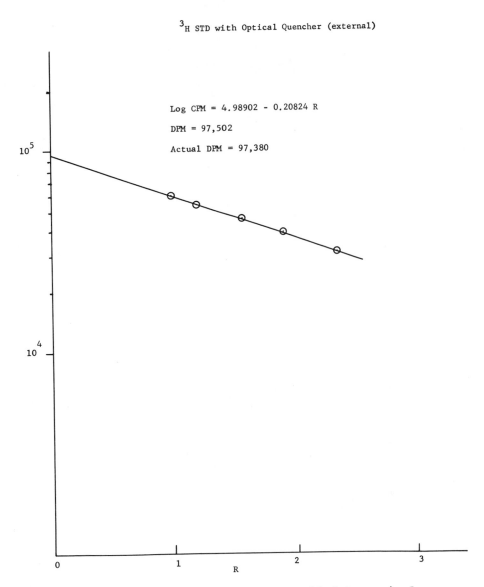

^{3}H STD with Optical Quencher (external)

Log CPM = 4.98902 - 0.20824 R

DPM = 97,502

Actual DPM = 97,380

Figure 4. Extrapolation method plot applied to a single sample containing ^{3}H but producing simulated quench by the use of optical filters around the counting sample vial.

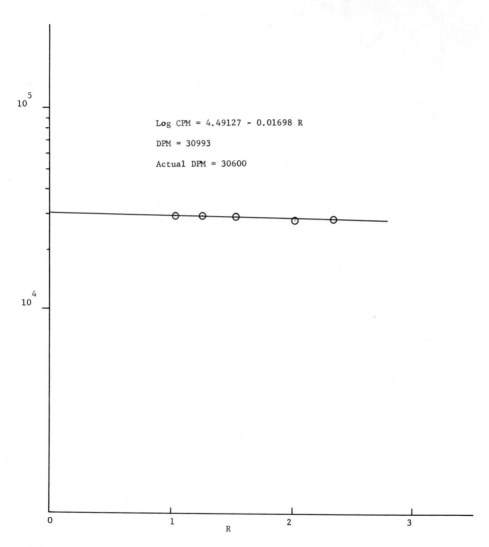

Figure 5. Extrapolation method plot applied to a single sample containing ^{14}C but producing simulated quench by the use of optical filters around the counting sample vial.

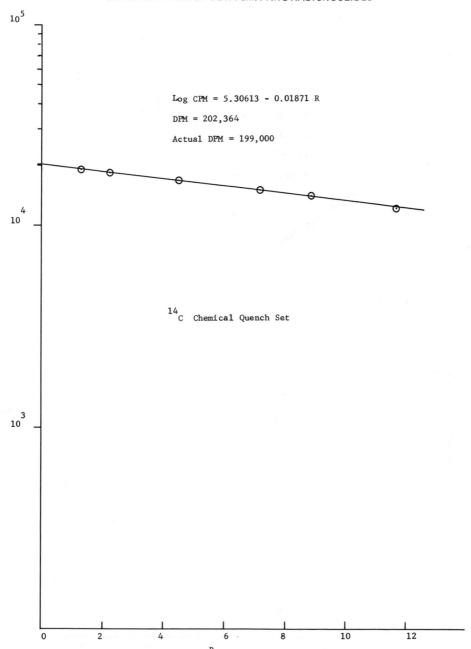

Log CPM = 5.30613 - 0.01871 R

DPM = 202,364

Actual DPM = 199,000

^{14}C Chemical Quench Set

Figure 6. Extrapolation method plot applied to a set of
samples with the same amount of ^{14}C but different
amounts of quench. Each R value corresponds to a
different counting sample.

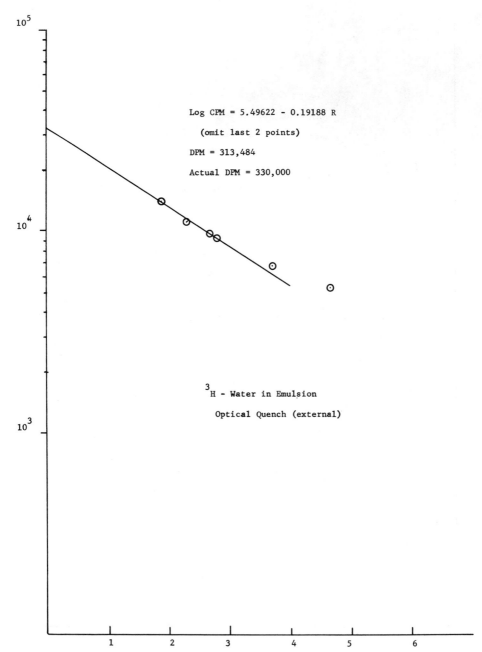

Figure 7. Extrapolation method plot applied to a single
sample containing ^{3}H-water but producing simulated
quench by the use of optical filters around the
counting sample vial.

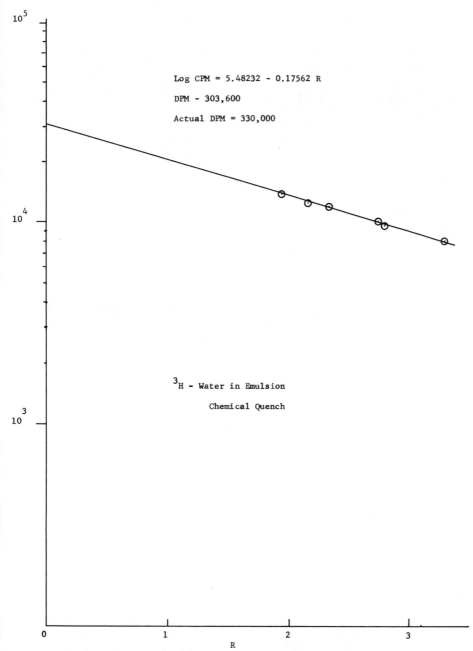

Figure 8. Extrapolation method plot applied to a set of samples with the same amount of ^{3}H-water but different amounts of quench. Each R value corresponds to a different counting sample.

CONCLUSIONS

The combinations of the H# and a coincidence type counter make possible the determination of the DPM of beta emitters by employing a single extrapolation. The method does have some limitations. It is not possible to start with highly quenched ^{3}H containing samples, i.e., R > 3.0. The method is probably accurate to within $\pm$ 3%. Further work will be performed to extend this method to a series of beta emitting nuclides and to further define the accuracy.

REFERENCES

1.   Horrocks, D.L., Nature (London) 202, 78 (1964).

2.   Horrocks, D.L., Progr. Nucl. Energy Sci. 9, 7, 21-110 (1966).

3.   Flynn, K.F., Glendenin, L.E., and Prodi, V., Organic Scintillators and Liquid Scintillation Counting, (D.L. Horrocks, and C.T. Peng, eds.), Academic Press, New York, 687 (1971).

4.   Horrocks, D.L., Patent pending.

5.   Horrocks, D.L., Nucl. Instr. Meth. 30, 157 (1964).

6.   Horrocks, D.L., Int. J. Appl. Radial. Isotopes 24, 49 (1973).

QUENCH CORRECTION CONSIDERATIONS IN HETEROGENOUS SYSTEMS

A.A. Noujaim, L.I. Wiebe, and C. Ediss
Division of Bionucleonics and Radiopharmacy
Faculty of Pharmacy and Pharmaceutical Sciences
University of Alberta
Edmonton, Alberta, Canada

Abstract

The behaviour of several commercial solubilizers
(PCS, Readysolv VI, Insta-Gel, and Aquasol) when quenched
with an aqueous system was examined. The relationship
between the external standard channels ratio (ESR) and the
isotope channels (C/R) ratio was determined before and after
phase change for both Tritium and Carbon-14. A comparison
was also made between the response of quasi-logarithmic
amplification and linear amplification operated in the
summed or lesser pulse-height mode. Observed data revealed
that neither the ESR nor C/R is a satisfactory method of
Quench Correction after phase separation. It does not
appear that the phase distribution of the Tritium radioactivity has any drastic effects on the results observed.

Introduction

The ever increasing use of commercial solubilizing
fluors has certainly simplified counting procedures of
aqueous samples. The use of such systems, however, could
result in the formation of non-homogenous liquids whose
scintillating characteristics are distinctly different
from the standard Toluene/PPO/POPOP mixture. The use of
counting solutions containing Triton X-100 has been
described over ten years ago (1,2). The effect of increasing
the percentage of water on the counting performance of such
systems has also been determined (3,4,5). In general, such
solutions are able to hold water as micelles, whereby the
water is held by the hydrophilic end of the Triton molecule.
The physical character of the micelle is a function of the
amount of water added as well as the temperature of the
environment. Reliability of automated quench correction
procedures is therefore subject to the above limitations.

It was the purpose of this investigation to compare the accuracy of the isotope channels ratio and the external standard ratio as applied to various commercial solubilizers.

Materials and Methods
    The following criteria were used in our evaluation:
(a)   The absolute counting efficiency as related to the sample channels ratio.
(b)   Comparison of the double ratio behaviour (6).
(c)   The figure of merit of the systems evaluated.
    The selection of instruments, instrument settings, and sample preparation was as follows:
Fluors: Commercial solubilizers known as Biosolv VI (Beckman Instruments Inc.), Insta-Gel (Packard Instruments Inc.), PCS (Amersham/Searle) and, Aquasol (New England Nuclear Corp.) were used throughout this investigation. A standard Bray's cocktail was also prepared (PPO 4 g, POPOP 0.2 g, Naphthalene 60 g, Ethyleneglycol 20 ml, Methanol 100 ml, and Dioxane 1000 ml).
Instrumentation: The response of a Picker Liquimat Model 220 exhibiting a quasi-logarithmic amplification system and variable windows was compared with that of a Searle Mark II (Searle Instrumentation) modified for both lesser and summed pulse height analysis (7). Window settings on this latter instrument are fixed by the manufacturer. The adjustment of the window openings on the Liquimat 220 was such that the best E^2/B was obtained for a channels ratio of $A/B = 0.3$ where channel A represents the narrow window and channel B represents the wide window. The unquenched standard used was Aquasol.
Sample Preparation: Ten millilitres of each fluor were spiked with known radioactivity of either ^{14}C-Toluene, ^{3}H-Toluene, or Tritiated water. Distilled water was used as a quenching agent, and incremental portions of 0.2 ml volumes were added to the original solution of fluor. An average of nine samples were used at each dilution level. All samples were dark adapted for thirty minutes after each dilution, and the gelling point was noted visually. Each sample was counted to a 1% error in the channel exhibiting least counts. All counting was performed at room temperature.

Results and Discussion
    We have ascertained that increasing the volume of the counting solution does not have any geometric effect on either the C/R or ESR. This is shown in Figure 1.

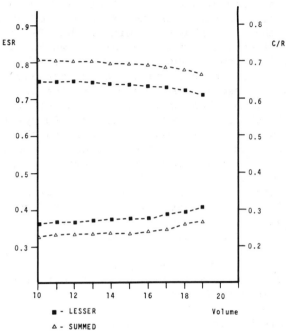

Figure 1.
Effect of volume on external standard channels ratio
and isotope channels ratio.
Increments of 1 ml Toluene added. (Picker Liquimat 220).

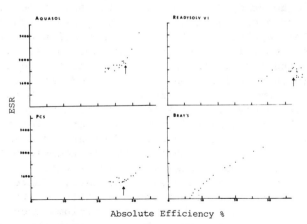

Figure 2.
Effect of water addition on Toluene-^{3}H ESR.
Arrows indicate visual phase change. (Picker Liquimat 220).

Thus, any observed changes could only be attributed to differences in pulse height distributions as a result of the addition of quenching agents. Figures 2 and 3 show the relationship between the absolute efficiency of tritium and Carbon-14 in the lipid phase and either the ESR or C/R. The three commercial fluors tested are known to suffer from phase separation upon the addition of a certain percentage of aqueous liquids. When compared to the standard Bray's cocktail, both Aquasol and PCS show a predictable relationship prior to visual phase change. Such a relationship becomes completely erratic after gelling. Readysolv VI on the other hand, exhibits only marginal change in the ESR either prior to or after phase changes. In the meantime, the absolute efficiency changes little prior to phase separation and considerably thereafter.

It was interesting for us to note that we have obtained essentially identical patterns with all four fluors when the Tritium was present in the aqueous phase rather than the lipid phase. The relative figures of merit are shown in Figure 4. The Biosolv VI preparation shows a serious deterioration of this value upon phase change. As expected, when the double ratio technique was applied to the systems investigated, (Figure 5) the unreliability of quench correction after visual phase change was evident. In the case of Biosolv VI, it was obvious that no correlation could be established, even when the sample appeared to be clear. A comparison of the response of lesser and summed pulse height analysis circuitries to the addition of water is illustrated in Figures 6 and 7. In this case, the double ratio accuracy deteriorates much rapidly before any visual phase change when the lesser pulse height mode is used. This applied equally to the situation where the Tritium was either in the lipid or aqueous phase. On the other hand, the lesser circuit proved to be far superior when C-14 Toluene in Insta-Gel was quenched with water (Figure 3). From our observations, it is then obvious that it is extremely dangerous to generalize as to a specific method of automated quench correction for all phase-combining systems. A number of other conclusions could also be drawn. For example, Bush's technique (6) of double-ratio method appears to be imperative in order to test the limits of any system to be used. Quite often, a specific quench correction technique fails before a phase change is visualized. In the case of Tritium, when present in either the aqueous of lipid phase, neither the C/R or

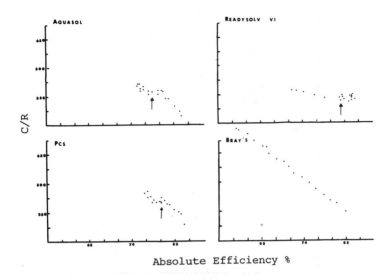

Figure 3.
Effect of water addition on Toluene C-14 C/R.
Arrows indicate visual phase change. (Picker Liquimat 220).

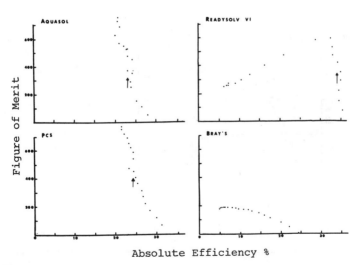

Figure 4.
Relationship between figure of merit (efficiency x % water added) and absolute efficiency.
Tritium present in aqueous phase in form of T_2O. (Picker Liquimat 220).

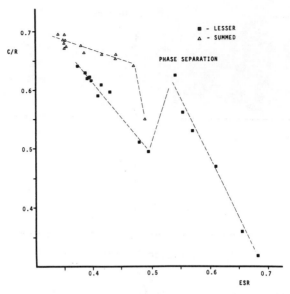

Figure 5.
Relationship between C/R and ESR for Toluene C-14.
Arrows indicate visual phase separation. (Picker Liquimat
220).

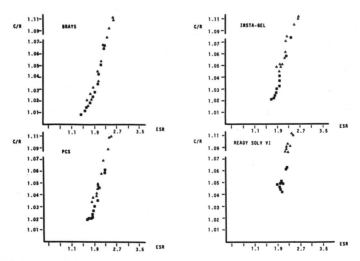

Figure 6.
Relationship between C/R and ESR.
Toluene-H^3 quenched by addition of water until visual
phase change.
$\triangle$ = summed PHA, ■ = lesser PHA. (Searle Mark II).

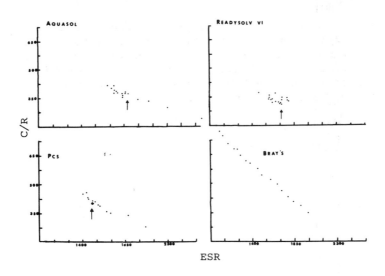

Figure 7.
Relationship between C/R and ESR.
Tritium present in aqueous phase in form of T_2O.
Δ = summed PHA, ■ = lesser PHA. (Searle Mark II).

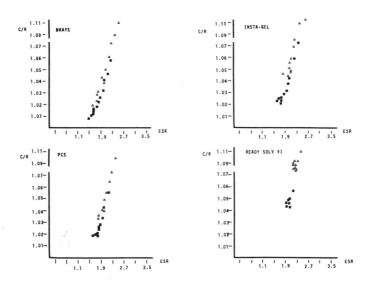

Figure 8.
Relationship between C/R and ESR.
Toluene C-14 in Insta-Gel. (Searle Mark II).

ESR is satisfactory after visual phase change. Internal
standards should then be used under such conditions. The
use of instruments with preset fixed windows in conjunction
with phase-combining cocktails could limit the method of
quench correction to the ESR technique, as the dynamic
range of the C/R is seriously reduced.

References
1. R.C. Meade and R.A. Stiglitz, Int. J. Appl. Rad. &
    Isotopes, 13, 11 (1962).
2. M.S. Patterson and R.C. Greene, Anal. Chem., 37,
    854 (1965).
3. J.C. Turner, Int. J. Appl. Rad. & Isotopes, 19,
    557 (1968).
4. D.A. Kalbhen and A. Rezvani, Organic Scintillators and
    Liquid Scintillation Counting, D.L. Horrocks, C.T. Peng,
    (Ed.), Academic Press, p. 149-167, 1971.
5. Y. Kobayashi and D.V. Maudsley, Liquid Scintillation
    Counting: Recent Developments, P.E. Stanley,
    B.A. Scoggins, (Ed.), Academic Press, p. 189-205, 1974.
6. E.T. Bush, Int. J. Appl. Rad. & Isotopes, 19, 447
    (1968).
7. C. Ediss, A.A. Noujaim, and L.I. Wiebe, Liquid
    Scintillation Counting: Recent Developments,
    P.E. Stanley, B.A. Scoggins (Ed.), Academic Press,
    p. 91-101, 1974.

THE USE OF THE LIQUID SCINTILLATION
SPECTROMETER IN BIOLUMINESCENCE ANALYSIS

Philip E. Stanley

Department of Clinical Pharmacology,
The Queen Elizabeth Hospital,
Woodville, South Australia 5011

ABSTRACT

This review covers publications concerning analytical
bioluminescence which in the main have appeared between
mid-1973 and mid-1976. Outlines of some new assays and
techniques are given together with modifications of
existing procedures. Comments are presented on the use
of the liquid scintillation spectrometer and other
equipment for measuring bioluminescence. New
applications are detailed and discussed.

INTRODUCTION

Light produced as a result of biological or enzymic
reactions and which arises from electronically excited
products is called bioluminescence. These reactions can
be used to advantage for the analysis of certain
compounds when they are present in rate limiting amounts.
Thus a sensitive assay for adenosine triphosphate (ATP)
and reduced nicotinamide adenine dinucleotide (NADH) can
be achieved using the luciferase enzyme-complex derived
from the firefly *Photinus* and the bacterium genus
Photobacterium respectively.

The present author reviewed extensively the
literature on this topic three years ago (1) and since

then other reviews have been published (2-5). This
review concerns in the main the literature appearing
between mid-1973 and May 1976 and only limited attention
will be paid to the basic principles since these have been
covered elsewhere (1-5). For up-to-date accounts of the
biochemistry of bioluminescence the reader is referred
to recent reviews (6,7).

GENERAL ASPECTS

Introduction

Analytical bioluminescence is based on measuring the
rate of the light emitting reaction. Thus it is
necessary to determine the number of photons emitted per
unit time at a fixed time or times after mixing the
reactants. The actual rate will be the observed rate
multiplied by an efficiency factor which must include not
only the quantum efficiency of the photomultiplier at the
wavelengths of the photons but also the optical efficiency
of the detector assembly. The quantum efficiency is
likely to be less than 10% for most photomultipliers (1)
and the optical efficiency (percentage of emitted photons
actually reaching the photocathode) is likely to be
between 25 and 75%. Thus the observed rate is of the
order 2 → 5% of the actual rate. The quantum yield of the
reaction is the average number of photons produced per
reacting molecule and for the firefly luciferase-ATP
system this value approaches one. For other bioluminescence reactions it is frequently a good deal less and
thus the worker usually only 'sees' a small fraction say
around 1% of the number of molecules undergoing the
bioluminescence reaction.

Temperature, pH, ionic strength etc. should be
rigorously controlled since these will influence the rate
of the reaction and consequently the final result of the
assay.

It is important to have information about the kinetics
of both the mixing of the reactants and the reaction
itself since there are distinct advantages in being able
to conduct an assay under conditions where the reaction
kinetics are slow relative to those of mixing, that is
when mixing is so fast that it has a negligible effect on

the reaction rate. This has been accomplished by Rhee *et al* (8) who used a stopped flow spectrophotometer to measure ATP with the firefly luciferase.

Another approach to measuring these reactions is to integrate the luminescence-time curve over a fixed interval commencing at the time of mixing or at a fixed time thereafter (9,10). This latter approach is adopted by most workers using the liquid scintillation spectrometer since photons are counted from the time the vial is loaded until the selected preset time, usually 5 → 30 seconds, has elapsed.

Yet another approach is to mix the reactants in a continuous stream and to pass the mixture through the detector in a flow system (11). Effluents from chromatography columns can be monitored in this manner. However, considerably more reagents are consumed and less information about the kinetics of the reaction can be gleaned since generally, measurements are made of the steady state situation.

INSTRUMENTATION

Commercial

The rationale for using the liquid scintillation spectrometer to measure bioluminescence has been dealt with extensively elsewhere (1,2,3,5,9,10,12) but suffice to say that it still is the most sensitive unit which is available in most laboratories which can perform this task. Single photons can be counted by switching the photomultipliers out of coincidence and adjusting the pulse height analyzer to count low energy pulses. A tritium channel is suitable. Care should be taken so as not to overload the analyzers with high count rates (>10^6 cpm) since a non-linear response may be evident (1,10). Some workers prefer to use one rather than both photomultipliers and choose that with the best efficiency and/or the lowest background. Other instrumentation is now available commercially for bioluminescence assays and these have been assessed and used by various workers. The units include:

1.   Lab-Line ATP-Photometer. Reagents are mixed in
a scintillation vial externally to the photometer and
then placed in it. Some 15 seconds after mixing, bioluminescence is integrated for one minute and the result
is then displayed digitally.

2.   DuPont Biometer. Reagents are mixed in a cuvette
adjacent to the photomultiplier. The sample (10 µℓ) is
introduced from a syringe through a septum and the total
reaction volume is limited to 100 µℓ. Mixing is
accomplished by the force of the injection and the efficacy
and reproducibility of this procedure must be checked for
the system under investigation. Bioluminescence is
integrated for 3 seconds after mixing and the result is
displayed digitally.

3.   Aminco Chem-Glow Photometer. Reagents are mixed
in a cuvette situated adjacent to the photomultiplier and
mixing again depends on the force of the injection of the
sample. The readout device is an analogue meter which can
be coupled to a recorder so it is possible to obtain data
over any required time interval. A flow cell is also
available for use in automatic analyses.

4.   Ortec - Brookdeal photon monitoring systems
fitted with lock-in amplifiers would appear to warrant
investigation for use in bioluminescence assays since they
provide excellent signal to noise ratio, pulse pair
resolution, maximum count capacity and are capable of very
high counting rates. This unit has been used in the
physical sciences but the author knows of no publications
in which this system has been used for bioluminescence
assays.

The silicon vidicon and vibrating mirror rapid scan
spectrometers can be used for measuring bioluminescence
spectra providing the light intensity is sufficient.

Recently two highly sensitive spectrophotofluorometers
have been described (13,14) which no doubt could be used
for measuring spectra from weak bioluminescence sources.

Some aspects of single photon counting

It has been shown (15) that the average count rate of
anode pulses from a photomultiplier

$$\bar{n} = I_a/Me = \gamma_a\Phi/Me = \gamma_c\Phi/e$$

where I_a = average anode current, Φ = light flux on the
cathode in watts, γ_a and γ_c are the sensitivities of the
anode and cathode respectively (A watts^{-1}), M is the
multiplication factor of the photomultiplier and e is the
charge of the electron. Thus $\bar{n}$ is proportional to Φ but
not to the applied high voltage which determines M and
this is a distinct advantage over the procedure involving
analogue signals. Photoemission is a statistical process
and given the signal to noise ratio is $\sqrt{\bar{n}t_c}$ (where t_c
is the counting time interval) it means that a lower
limit for a measurable rate can be established for a given
system. The maximum count rate is also limited by this
statistical process and it is recommended that the maximum
count rate of the system (non-random pulses) exceed $\bar{n}$
(random pulses) by at least a factor of twenty so as to
avoid loosing counts during times when the instantaneous
rate exceeds this maximum. These workers also point out
that the input time-constant (RC) of the photomultiplier
will influence the height and width of the photoelectron
pulses and ultimately of course limit its maximum counting
rate.

The advent of very fast phase sensitive amplifiers will
no doubt play an important part in single photon counting
since in principle at least, a proportion of photomultiplier background pulses can be eliminated from true
signal pulses just on the basis of their shape.

Special instruments

Beall and Haug (16) have designed an instrument to
measure the very fast kinetics associated with the light
emission of the photosynthetic alga *Scenedesmus* for a
20 μsec period after a stimulus light was switched off.
Such a system may be valuable for studying the kinetics of
bioluminescence reactions although a recent report (17)
suggests at least for the kinetics of firefly luciferase,
such extreme measuring speeds are unnecessary since after
mixing ATP, luciferin and luciferase there is a delay of
some 25 msec before light is emitted and the maximum

intensity occurs after 300 msec.

Wettermark *et al* (18) have designed a photon counting
device coupled to a multichannel analyzer operated in
multiscale to measure ATP in single cells using the firefly
luciferase system. A few femtomoles (10^{-15} moles) of ATP
could be readily measured.

Rhee *et al* (8) have measured ATP in the picomole range
using a stopped flow spectrophotometer and have concluded
that the best index of ATP concentration is obtained by
measuring the rate of the initial rise in bioluminescence
following mixing. A similar conclusion has been reached
by Lundin and Thore (19) who used a specially designed
photometer fitted with an automatic dispenser/injector
coupled to an electronic timer. These workers also studied
the influence of injection velocity and hence the kinetics
of mixing on the initial phase of bioluminescence and peak
height reached.

Chappelle and co-workers have published details of the
detection of microbial cells in urine and the automatic
instruments they have employed (20-22) while Allen has
described the development of a luminescence biometer for
detecting microbial cells (23).

Automatic or semi-automatic instruments will no doubt
be seen in increasing numbers where large sample numbers
are involved such as for monitoring the microbiological
disposal of sewage and in measuring biomass in soils and
marine and fresh waters.

Automatic injection

To deal with large numbers of samples using a liquid
scintillation spectrometer, some degree of automation is
essential for mixing the reactants. Reproducible mixing
is most important especially if the kinetics of the
reaction is of the same order as that of mixing. While one
kind of mixing may be satisfactory for one assay it may not
be at all adequate for another especially if the densities
of reactant solutions are substantially different or the
solutions are viscous. Often workers and instruments rely
on the forcible injection of a reactant to cause mixing.
Stirring of reactants to cause mixing is an alternative but
in the automatic LSC stirring is almost impossible but in

the photometer this can be achieved, however, completely
light tight seals around the stirrer are difficult to make.
Magnetic stirring should be avoided since this may defocus
the photomultipliers unless there is an adequate mu-metal
shield.

Hammerstedt (24) has interfaced a Hamilton precision
liquid dispenser to a liquid scintillation counter so that
in the assay of ATP, the firefly luciferase is added to
the next sample just as the data from the previous sample
is being printed. Consistant and precision timing of the
mixing process is thus accomplished. More recently a
pneumatically operated manual dispenser has been described
which mixes 10 μℓ ATP solution into 100 μℓ of the more
viscous luciferase preparation (25). Samples were then
counted in the spectrometer. Several types of dispenser
were evaluated and the workers report that mixing which is
too vigorous causes the enzyme to luminescence even in the
absence of ATP. Reproducible injection of ATP solutions
into firefly luciferase enzyme system has been described
by Brunker (26). Here the injection of ATP is made from a
modified syringe fitted into a scintillation vial. The
syringe plunger is operated automatically by a hinge
actuated by the light tight shutter of the detector of the
liquid scintillation spectrometer.

ADVANTAGES

The advantages of bioluminescence assays or analytical
bioluminescence include high sensitivity, specificity,
speed and economy. Sensitivity will depend ultimately on
the photomultiplier used; its thermal noise and quantum
efficiency at the wavelength of interest being most
important (1). The quantum efficiency of the reaction is
of course very significant. Thus in the ATP-firefly
luciferase system the value approaches unity. This is many
orders of magnitude greater than the values recorded in
chemiluminescence. It is of interest that ATP itself has
been shown to be chemiluminescent with a quantum efficiency
of 10^{-16} - 10^{-17} at pH 7 - 7.5 (27). Specificity for
bioluminescence reactions is often high as is the case in
the ATP-firefly luciferase system. No other naturally
occurring nucleotide triphosphate is effective (however
3-iso-AMP and ε-AMP are active; see (1)). Bacterial
luciferases can be used to measure NADH (10,12) but the

sensitivity of NADPH is twenty times less. Flavin mono-
nucleotide is also required and thus can be measured in
this reaction (10,12).

Specificity is high in the bioluminescent system of
Renilla since only PAP (adenosine 3'-phosphate 5'-
phosphate) is active. Thus the assay of PAP and PAPS
(adenosine 3'-phosphate 5'-sulphatophosphate) using the
LSC has been described (1,28,29,30). The occurrence of
PAPS has been unequivocally established in plant materials
using this method (30).

Bioluminescence reactions can often be performed
quickly since they require no separation procedures such
as are usually required for example with radioisotopic
techniques. They are thus similar in many respects to
spectrophotometric assays used for instance to follow the
formation of NADH. The main difference to the spectrophotometric procedure is of course sensitivity, the bioluminescence assay being at least 20,000 times more
sensitive. The assay of malate, oxalacetate, pyrophosphate,
adenosine 5'-sulphatophosphate, PAPS, takes around seven
or eight minutes to perform (1).

The cost of the assays is small since the reagents
usually cost only a few cents per assay. Where large
numbers of samples are concerned this becomes important as
is the case in the field of clinical biochemistry for
example in measuring various dehydrogenase enzymes or for
screening the blood of newborn infants for creatine
phosphokinase to detect muscular dystrophy (31).

NEW AND MODIFIED ASSAY PROCEDURES

Of the recent publications most are concerned in some
way with ATP. Little about the assay of NADH and its
conjugates has appeared.

Cheer *et al* (32) have described a procedure to measure
ATP using a liquid scintillation spectrometer and it was
similar to previously described methods except that
counting was performed with the photomultipliers in
coincidence. The potential problems associated with this
approach have been discussed elsewhere (9,10,12). As would
be expected the enzyme background was much reduced when

216

compared to that obtained with the out-of-coincidence technique. These authors provided good evidence that samples containing ATP were best kept at the temperature of liquid nitrogen or dry ice since at -20^0C considerable losses of ATP were apparent. Kimmich *et al* (33) also used the photomultipliers in coincidence because of high and variable enzyme blanks. These workers have devised a method for measuring both AMP and ADP after converting them enzymically to ATP and in addition have achieved a five-fold activation of the crude extract of firefly luciferase by treating it with calcium phosphate. Further they employed a system buffered at a pH of 8.0 instead of the usual pH 7.4 and obtained an assay with an enhanced sensitivity.

Kimmich *et al* (33), Weiner *et al* (34) and Stanley (1,9) have recognized that ATP coprecipitates with potassium perchlorate formed when perchloric acid extracts are neutralized with KOH. If possible this neutralization step should be avoided and the extract assayed directly using an adequately buffered system (9). Davison and Fynn (35) however recommended that the sample be extracted into perchloric acid and then centrifuged to remove all protein since they have found in *Bacillus brevis* an ATPase which was not denatured by the acid. Without the centrifuging procedure the enzyme would be carried over at the subsequent neutralization and other steps and so hydrolyse ATP in the sample.

Manandhar and van Dyke have used the firefly luciferase system to measure adenosine tetraphosphate (36,37), guanosine triphosphate (37,38) and both purine and pyrimidine ribose and deoxyribose nucleoside triphosphates (39). The light output takes several minutes to reach a maximum and it seems likely that such kinetics are due to the various phosphates being converted to ATP by enzymes present in the crude firefly lantern extract. The authors give details of the thin layer chromatography necessary to separate the nucleotides prior to analysis (36,38).

The continued interest in the bioluminescence assay for cyclic nucleotides is shown by the publication of another sensitive assay for adenosine 3',5'-monophosphate (40) and another for the assay of guanosine 3',5'-monophosphate phosphodiesterase activity (41). Enzymic conversion to

ATP is the basis of both procedures.

A multichannel analyzer used in the multiscale mode is used by Quammen *et al* (42) to follow the kinetics of luminescence decay in the assay of ATP using a liquid scintillation spectrometer. In addition a flow type cell was employed so that the reactants could be mixed in the detector assembly rather than in the laboratory. The present author has followed similar kinetics using the same approach (1,28).

A recent publication (43) has highlighted a problem often not appreciated by workers new to the field and which tends to discourage them from exploiting analytical bioluminescence to its full extent. This is the problem of phosphorescence in glass and polythene vials. It is often caused by their exposure to fluorescent or direct sunlight. This spurious light is similar to bioluminescence in that it is composed of single photons and cannot be discriminated from it electronically. However it usually decays to negligible values within one to five minutes. Corredor *et al* (43) has shown the decay rate is different for various brands of vials presumably because different kinds or batches of glass and polythene were used in their manufacture. Further the light transmission (bioluminescence) can vary by a factor of more than 1.6 while the within vial variation of each type was 5 → 10%. Polythene vials gave the best transmission but also the highest phosphorescence, sometimes so high as to preclude their use.

A very sensitive assay for proteolytic enzymes has been described by Njus *et al* in which bacterial luciferase is used as the substrate (44). This luciferase enzyme has as its substrate FMN and not ATP. The luciferase was treated with the protease separately and the remaining active luciferase was measured at various times in a standard assay procedure. As little as 20 ng trypsin could be measured and no separation procedures were involved. The present author has shown that the assay can be followed in a dynamic fashion using the liquid scintillation spectrometer (set in the repeat count mode) when trypsin is mixed with the bacterial dehydrogenase/luciferase and with NADH and FMN (11) present in excess. Presumably trypsin acts on both the dehydrogenase as well as the luciferase.

Nakamura (45) has standardized a light source containing luminol and has used it to measure quantum yields of bioluminescent reactions. Such a standard system should be considered by workers who quote the sensitivity of their procedures. Inter-laboratory and inter-instrument comparisons could be therefore readily made.

APPLICATIONS

Measurement of Biomass

In recent years ATP has shown great promise as an index of biomass and the pioneering work of Holm-Hansen and colleagues has been discussed previously (1). It is recognised that the efficient extraction of intact ATP from the microorganism in the water or soil is the most difficult and critical step. Perchloric, trichloroacetic and sulphuric acids, boiling water, boiling buffers including tris, tris-borate and glycine, dimethylsulphoxide, n-butanol, n-bromosuccinamide and boiling chloroform have been used (1,46). Soils and sediments are particularly difficult to deal with and it appears that procedures involving cold acid extraction are the methods of choice. Karl and LaRock (46) used sulphuric acid-EDTA and obtained extraction of 81-94% whereas with boiling tris the recovery was 3-6%. The selection of an appropriate internal standard is difficult since native ATP will be extracted much more readily for example than ATP derived from a microorganism adhering tenaciously to a particle of soil. Lyophilized bacteria may be employed or bacterial cells coated onto glass beads, similar in size to the soil particles, may be suitable. These of course should be added prior to the extraction process. Karl and LaRock (46) stress the importance of the effect of extracted cations and anions on the luciferase enzyme.

Moriarty (47) has recently shown that there is a correlation between muramic acid and biomass of bacteria in aquatic sediments. Since Gram-negative bacteria contain about 20 µg muramic acid/mg C and Gram-positives have 100 µg/mg C the relative numbers of each type must be estimated by other means. The method cannot be used in presence of blue green alga. Other publications on measuring biomass by the ATP index are to be found in the references (48-53).

Other applications of the firefly luciferase ATP assay

This procedure has been used extensively to study ATP in various components of blood. It is considered that the concentration of adenine nucleotides have critical influence on platelet aggregation *in vitro* and David and Herion (54) have devised and tested a technique for studying both ADP and ATP levels in such a system. The biolumines- cence assay has also played its role in investigating the kinetics of thrombin-induced release of ATP by platelets (55). The level of ATP has been measured in red cells (56,57) and both groups of workers showed that the bioluminescence procedure gave falsely high readings due to luciferase being stimulated by a protein present in the extracts. The latter report (57) showed that denatured haemoglobin increases the light output.

The mechanism for the discocyte-echinocyte shape transformation in normal and ATP-enriched human erythro- cytes has been studied using the firefly luciferase-ATP assay (58) and depletion or repletion of ATP in conjunction with intracellular or intramembrane factors has been suggested as being responsible for the equilibrium. Decreased levels of ATP in erythrocytes have been detected using the firefly luciferase-ATP procedure. Thus Wolf *et al* (59) have observed such a decrease in patients with either sickle cell anaemia, alcoholic cirrhosis, viral hepatitis or chronic renal disease.

The release or efflux of ATP from biological tissues (60) and from perfused heart during coronary vasodilation (61) have also been reported. The luciferase-ATP assay has also been used to study the effect on ATP levels of ultra high frequency radiation (2.45 GHz at 50 mwatts/cm^2) on liver hepatocytes (62) and to measure ATP in Tarantula spider venoms. In the latter the concentration of ATP was surprisingly high (28-57 µg/µℓ venom) and it was shown to act synergistically with the venom toxin (63).

The luciferase-ATP method has also been used to effect in plant biochemistry to study the phytochrome mediated changes in ATP concentration in bean buds (64,65) and the light induced concentration changes of ATP in sporangio- phores of *Phycomyces* (66). In a study of ATP formation in the mitochondria of etiolated corn shoots treated with herbicide Rushness and Still (67) observed that the firefly luciferase assay was inhibited by the isopropyl-3-chlorocarbonilate and two of its hydroxylated metabolites. This observation indicates just how important it is to include the correct controls when bioluminescence assays are being used.

Considerable interest has centred on measuring ATP in bacterial cells (1) and this trend continues from both the research and applied side. Thus ATP pools in *Nitrobacter* have been shown to be as high as 8 picomoles per μg cell N (68) and Strange *et al* (69) in an older study investigated the effect of starvation on the ATP concentration in *Aerobacter* while Kao *et al* (70) have measured the pools in *E. coli* and *Pseudomonas*. The detection of foreign or contaminating microorganisms is pertinent here and Sharpe *et al* (71) have used the luciferase-ATP assay to detect microorganisms in various foodstuffs and found that during the incubation of these foodstuffs, the intrinsic ATP decreased while microbial ATP increased. In the clinical field the luciferase-ATP assay is being developed for the detection of bacteruria (72,73). ATP from mammalian cells which are present in the urine, was extracted by treatment with Triton X-100 and then hydrolysed with apyrase. The microbial ATP was then estimated after its extraction with boiling tris-buffer or perchloric acid. The lower limit of detection reported by both groups was around 10^5 cells. It has the advantage of speed and there is the potential for screening the sensitivity of organisms to various therapeutic agents.

The on-line measurement of the biosynthesis of ATP has been covered extensively elsewhere (1) and the procedure has been used recently to study the ATP produced in mitochondrial suspensions (74) and ATP release from platelets (55). The rate of ATP synthesis in bovine anterior pituitary slices has also been measured (75) and a study has been made of three iso-enzymes of human creatine phosphokinase and their levels following myocardial infarction (76). A sensitive enzyme assay for reverse reaction of the first histidine biosynthetic enzyme has been described recently (77). Thus low levels of ATP-phosphoribosyl transferase has been measured by estimating the ATP released from the enzyme's substrate, N'-(5'-phospho-D-ribosyl)-ATP.

Surprisingly little has appeared about the use of the
NADH-bacterial luciferase assay although the present
author has found it useful for measuring alcohol
dehydrogenase and other dehydrogenases. Recently however,
Hammar (78) has used bacterial luciferase to measure the
epidermal activity of NAD-dependant iso-citrate
dehydrogenase in skin biopsies during treatment with
dithranol.

Izutsu *et al* (79) have studied a bioluminescence assay
for ionic calcium which is based on its interaction with
the protein aequorin derived from the jellyfish *Aequorea*.
To obtain good reproducibility of results they found rapid
mixing was necessary and concluded that the accuracy of
the assay depended on the association constants of the
calcium chelating agents in the test solution.

Kinetic studies have been made of the effect of
various anaesthetics such as halothane and fluroxene on
cell-free preparations of firefly lanterns (80,81) and the
evidence for the observed inhibition of light output
suggests that the anaesthetic acts at a hydrophobic site
on the luciferase molecule causing a structural alteration
and thus change in activity of the enzyme.

Luminescent bacteria continue to prove useful models
for the study of certain anaesthetics since the levels
inhibiting 50% bacterial luminescence and general
anaesthesia in mammals is remarkably similar (82).
Chloroform has been shown to be a competitive inhibitor of
dodecanal in the light producing reaction (83). Two recent
papers describe the effect of diethyl ether on the *in vivo*
and *in vitro* light emission from *Vibrio fischeri* (84,85).

CONCLUSION

Analytical bioluminescence continues to be used by a
gradually increasing number of workers and for a widening
range of applications. Their low cost and high sensitivity
make them particularly attractive for clinical biochemistry
and also as enzyme labels or means of assay in the field
of enzyme-**immunoassay.** The other area of potential is in
the biochemistry of small numbers of cells as is the case
for some tissue cultures and amniocenteses. The bacterial
enzyme is particularly attractive for routine work since
it can be obtained from readily grown cultures.

ACKNOWLEDGEMENTS

I thank Mrs. Ermioni Mourtzios for her patience and careful typing of this manuscript.

REFERENCES

1. P.E. Stanley in Liquid Scintillation Counting, Vol. 3, p. 253 (M.A. Crook and P. Johnson, Eds.). London and New York : Heyden & Son (1974).

2. E. Schram in Liquid Scintillation Counting : Recent Developments, p. 383 (P.E. Stanley, B.A. Scoggins, Eds.). New York : Academic Press (1974).

3. E. Schram, Arch. Int. Physiol. Biochem. 81, 561 (1973)

4. W.R. Seitz and M.P. Neary, Anal. Chem. 46, 188A (1974).

5. K. van Dyke, Packard Technical Bulletin, Number 20, November 1974, Packard Instrument Company Inc., Downers Grove, Ill. U.S.A.

6. M.J. Cormier, J. Lee and J.E. Wampler, Ann. Rev. Biochem. 44, 255 (1975).

7. J.W. Hastings, Ciba Foundation Symposium 31, 125 (1975)

8. S.G. Rhee, M.I. Greifner and P.B. Chock, Anal. Biochem. 66, 259 (1975).

9. P.E. Stanley, and S.G. Williams, Anal. Biochem. 29, 381 (1969).

10. P.E. Stanley, Anal. Biochem. 39, 441 (1971).

11. W.R. Seitz and D.M. Hercules in Chemiluminescence and Bioluminescence, p. 427 (M.J. Cormier, D.M. Hercules and J. Lee, Eds.). New York : Plenum (1973).

12. P.E. Stanley in Organic Scintillators and Liquid Scintillation Counting, p. 607 (D.L. Horrocks and C.-T. Peng, Eds.). New York : Academic Press (1971).

13. P. Vigny and M. Duquesne, Photochem. Photobiol. 20, 15 (1974).

14. S. Cova, G. Prenna and G. Mazzini, Histochem. J. 6, 279 (1974).

15. Appendix in Luminescence Spectrometry in Analytical Chemistry, p. 340 (J.D. Winefordner, S.G. Shulman and T.C. O'Haver, Eds.). New York and London : Wiley-Interscience (1972).

16. H.C. Beall and A. Haug, Anal. Biochem. 53, 98 (1973).

17. M. DeLuca and W.D. McElroy, Biochemistry 13, 98 (1974).

18. G. Wettermark, H. Stymne, S.E. Brolin and B. Petersson, Anal. Biochem. 63, 293 (1975).

19. A. Lundin and A. Thore, Anal. Biochem. 66, 47 (1975).

20. E.W. Chappelle and G.V. Levin, Navy Contractor Report 178-8097 (1964).

21. G.L. Picciolo, B.N. Kelbaugh and E.W. Chappelle, NASA GSFC Document X-641-71-163, (1971).

22. B.N. Kelbaugh, G.L. Picciolo and E.W. Chappelle, U.S. Patent 3,756,920, September 4, 1973.

23. P.D. Allen, Devel. Indust. Micro. 14, 67 (1973).

24. R.H. Hammerstedt, Anal. Biochem. 52, 441 (1971).

25. R. Johnson, J.H. Gentile and S. Cheer, Anal. Biochem. 60, 115 (1974).

26. R.L. Brunker, Anal. Biochem. 63, 418 (1975).

27. T.N. Livanova, Byull. Eksp. Biol. Med., 78, 39 (1974).

28. P.E. Stanley in Liquid Scintillation Counting: Recent Developments, p. 421 (P.E. Stanley and B.A. Scoggins, Eds.). New York and London : Academic Press (1974).

29. P.E. Stanley, in Chemiluminescence and Bioluminescence p. 494 (M.J. Cormier, D.M. Hercules and J. Lee, Eds.). New York : Plenum (1973).

30. P.E. Stanley, B.C. Kelley, O.H. Tuovinen and D.J.D. Nicholas, Anal. Biochem. 67, 540 (1975).

31. H. Zellweger and A. Antonik, Pediatrics 55, 30 (1975).

32. S. Cheer, J.H. Gentile and C.S. Hegre, Anal. Biochem. 60, 102 (1974).

33. G.A. Kimmich, J. Randles and J.S. Brand, Anal. Biochem. 69, 187 (1975).

34. S. Wiener, R. Wiener, M. Urivetzky and E. Meilman, Anal. Biochem. 59, 489 (1974).

35. J.A. Davison and G.H. Fynn, Anal. Biochem. 58, 632 (1974).

36. M.S.P. Manandhar and K. van Dyke, Anal. Biochem. 58, 368 (1974).

37. M.S.P. Manandhar, K. van Dyke and R.L. Robinson, The Pharmacologist 15, 205 (1973).

38. M.S.P. Manandhar and K. van Dyke, Anal. Biochem. 60, 122 (1974).

39. M.S.P. Manandhar and K. van Dyke, Microchem. J. 19, 42 (1974).

40. D. Glick, Y. Katsumata and D. von Redlich, J. Histochem. Cytochem. 22, 395 (1974).

41. R. Fertel and B. Weiss, Anal. Biochem. 59, 386 (1974).

42. M.L. Quammen, P.A. LaRock and J.A. Calder, in Estuarine Microbial Ecology, p. 329 (L.H. Stevenson and R.R. Calwell, Eds.). South Carolina : University of South Carolina Press (1973).

43. J.E. Corredor, D.G. Capone and K.E. Cooksey, Anal. Biochem. 70, 624 (1976).

44. D. Njus, T.O. Baldwin and J.W. Hastings, Anal. Biochem. 61, 280 (1974).

45. T. Nakamura, J. Biochem. (Tokyo) 72, 173 (1972).

46. D.M. Karl and P.A. LaRock, J. Fish. Res. Board Can. 32, 599 (1975).

47. D.J.W. Moriarty, Oecologia (Berl.) 20, 219 (1975).

48. W. Ernst, Oecologia (Berl.) 5, 56 (1970).

49. J.W.M. Rudd and R.D. Hamilton, J. Fish. Res. Board Can. 30 , 1537 (1973).

50. R.L. Ferguson and M.B. Murdoch in Nat. Mar. Fish. Ser. Ann. Report to A.E.C. (1973).

51. R.R. Christian, K. Bancroft and W.J. Weibe, Soil Science 119, 89 (1975).

52. A.D. Jassby, Limnol. Oceanogr. 20, 646 (1975).

53. G.V. Levin, J.R. Schrot and W.C. Hess, Environ. Sci. Technol. 9, 961 (1975).

54. J.L. David and F. Herion in Advances in Experimental Medicine and Biology, Vol. 34, p. 341 (P.M. Mannucci and G. Gorin, Eds.). New York : Plenum Press (1972).

55. T.C. Detwiler and R.D. Feinman, Biochemistry 12, 2462 (1973).

56. E. Beutler and C.K. Mathai, Blood 30, 311 (1967).

57. G.J. Brewer and C.A. Knutsen, Clin. Chim. Acta 14, 836 (1966).

58. C.J. Feo and P.F. Leblond, Blood 44, 639 (1974).

59. P.L. Wolf, P. Walters and P. Singh, Fed. Proc. 35, 252 (1976).

60. E.M. Silinsky, Comp. Biochem. Physiol. A 48, 561 (1974).

61. B.M. Paddle and G. Burnstock, Blood Vessels 11, 110 (1974).

62. E.N. Albert, G. McCullars and M. Shore, J. Microwave Power 9, 205 (1974).

63. T.K. Chan, C.R. Geren, D.E. Howell and G.V. Odell, Toxicon 13, 61 (1975).

64. J.M. White and C.S. Pike, Plant Physiol. 53, 76 (1974).

65. R.L. Kirshner, J.M. White and C.S. Pike, Physiol. Plant 34, 373 (1975).

66. W. Shropshire Jr. and K. Bergman, Plant Physiol. 43, 1317 (1968).

67. D.G. Rushness and G.G. Still, Pest. Biochem. Physiol. 4, 109 (1974).

68. U. Eigener, Archiv. Microbiol. 102, 233 (1975).

69. R.E. Strange, H.E. Wade and F.A. Dark, Nature 199, 55 (1963).

70. I.C. Kao, S.Y. Chiu, L.T. Fan and L.E. Erickson, J. Water Poll. Control Fed. 45, 926 (1973).

71. A.N. Sharpe, M.N. Woodrow and A.K. Jackson, J. Appl. Bact. 33, 758 (1970).

72. A. Thore, A. Lundin and S. Bergman, J. Clin Microbiol. 1, 1 (1975).

73. R.B. Conn, P. Charache and E.W. Chappelle, Amer. J. Clin. Path. 63, 493 (1975).

74. J.J. Lemasters and C.R. Hackenbrock, Biochem. Biophys. Res. Commun. 55, 1262 (1973).

75. P. Sheterline and J.G. Schofield, Biochim. Biophys. Acta 338, 505 (1974).

76. S.A.G.J. Witteveen, B.E. Sobel and M. DeLuca, Proc. Nat. Acad. Sci. U.S.A. 71, 1384 (1974).

77. J. Kleeman and S.M. Parsons, Anal. Biochem. 68, 236 (1975).

78. H. Hammar, J. Invest. Dermatol. 65, 315 (1975).

79. K.T. Izutsu, S.P. Felton, I.A. Siegel, J.I. Nicholls, J. Crawford, J. McGough and W.T. Yoda, Anal. Biochem. 58, 479 (1974).

80. I. Ueda, Anesthesiology 26, 603 (1965).

81. I. Ueda and H. Kamoya, Anesthesiology 38, 425 (1973).

82. M.J. Halsey and E.B. Smith, Nature 227, 1363 (1970).

83. G. Adey, C.R. Dundas and D.C. White, Brit. J. Anaesth. 45, 643 (1973).

84. A.J. Middleton and E.B. Smith, Proc. Roy. Soc. London Ser. B. 193, 159 (1976).

85. A.J. Middleton and E.B. Smith, Proc. Roy. Soc. London Ser. B. 193, 173 (1976).

QUANTITATION OF LEUKOCYTE CHEMILUMINESCENCE FOLLOWING
PHAGOCYTOSIS: TECHNICAL CONSIDERATIONS USING
LIQUID SCINTILLATION SPECTROMETRY

D. English*, A.A. Noujaim**, T. Horan*, and T.A. McPherson*
*Department of Medicine,
Dr. W.W. Cross Cancer Institute
**Faculty of Pharmacy and Pharmaceutical Sciences,
University of Alberta
Edmonton, Alberta, Canada

Abstract
    Electron excitation of oxygen molecules, with release
of photons due to electron relaxation, is associated with
leukocyte phagocytosis. These events are the result of
formation of singlet oxygen and/or superoxide anion, and
these, in turn, are unique intracellular microbicidal
agents. Rate of photon emission can be monitored in a
liquid scintillation counter operated at 25-27°C in the
out-of-coincidence mode. Both phagocytic monocytes and
polymorphonuclear leukocytes isolated from human blood
emitted photons upon engulfment of opsonized zymosan, or
yeast. Rate of photon emission was proportional to the
concentration of phagocytes present in the sample. Color
quenching of photons was noted when contaminating erythrocytes were present in the sample. For this reason, a
technique was developed to prepare highly purified preparations of mononuclear and polymorphonuclear leukocytes
devoid of erythrocytes. Control studies indicated that
chemiluminescence was inhibited if Ca^{++} and Mg^{++} were
absent from the test sample. Chemiluminescence appears to
be a simple method for determining phagocytosis and
intracellular killing on the part of phagocytic monocytes
and polymorphonuclear leukocytes present in human blood.

Introduction
    Human polymorphonuclear leukocytes (PMNL) bioluminesce
upon phagocytosis of bacteria (1). This bioluminescence
may be due to the generation of electron excitation states
associated with the generation of singlet oxygen ($^1O_2^{\cdot}$) or

229

to the generation of superoxide anion (O_2^-; 1, 2). On
the other hand, the molecular explanation for bioluminescence
is thought to be unknown by some investigators (3), and there
is a great deal of uncertainty as to the relationship
between the chemiluminescence and the bactericidal activity
of PMNL (3). PMNL from patients with chronic granulomatous
disease of childhood fail to produce superoxide anion or
hydrogen peroxide after phagocytosis (4). This is associated
with a deficiency of bacterial killing (4) and a lack of
emission of photons after phagocytosis (2).

A previous report (5) indicates that the emission
photon wavelength can be observed as a wide peak of photon
emission in the visible region. Singlet oxygen is known
to emit photons at distinct peaks of 7030, 6340, 5200 and
4800 Å (6). Studies from our laboratory (unpublished
results) suggest that two distinct peaks of photon emission
at 4360 and 5500 Å are associated with the chemiluminescence
of phagocytosis using human PMNL. These peaks are of a
higher energy than the major peak resulting from the
relaxation of singlet oxygen, which is 6340 Å (6). Thus,
it is unlikely that singlet oxygen is directly responsible
for chemiluminescence, although its role as an indirect
factor cannot be excluded.

PMNL bactericidal effects are dependent on a halide
cofactor (3), and chemiluminescence by phagocytes is also
dependent on a halide cofactor (7). Mixture of hydrogen
peroxide with hypochlorite, produced by PMNL phagocytosis,
is associated with a weak red chemiluminescence thought
to be due to the generation of singlet oxygen. This production of chemiluminescence appears to be dependent on
the presence of a halide cofactor (Cl^-). Further, the
singlet oxygen quencher, 1,4-diazobicyclo [2,2,2] octane
(8) inhibits the myeloperoxidase-mediated bactericidal
system of the phagocyte (9). Finally, superoxide dismutase
will inhibit both chemiluminescence and the bactericidal
effects of PMNL phagocytes (2). It has been suggested
that superoxide dismutase is present in most mammalian
cells to protect them from the toxic effects of radicals
produced by the univalent reduction of oxygen (10), but
may be compartmentalized in the PMNL so as to provide the
PMNL with a unique and potent bactericidal mechanism (11).

Thus, chemiluminescence associated with phagocytosis
may be due to the activation of the hydrogen peroxide-
halide-myeloperoxidase bactericidal system. Detection and
quantitation of such chemiluminescence might therefore be
used as a test of phagocytic function. This report

describes the technical considerations involved in the
quantitation of PMNL chemiluminescence using liquid
scintillation spectrometry.

Materials and Methods
 Liquid Scintillation Counter: A Searle Instrumentation
Isocap 300 Scintillation Counter was used for most of the
experiments in this study. The counter was operated at
ambient temperature (25-27°C) with the coincidence gate
bypassed so as to record all photons exciting a single
photomultiplier tube. A wide window was used to record all
counts.
 Vials: Kimble low potassium glass scintillation vials
were used throughout the study. Vials were kept in the
dark prior to use and prescreened for background chemiluminescence prior to each experiment. A single-tube
background of <10,000 CPM was considered acceptable.
 Leukocyte Preparations: Four different preparations
of leukocytes were used in the study (Figure 1). Aseptic
siliconized glassware was used throughout.
 A mixed preparation of leukocytes and erythrocytes
was isolated from heparinized venous blood (10 ml), by
diluting the blood to 20% with plasmagel (HTI Laboratories,
Buffalo, New York) to increase erythrocyte sedimentation.
This was held at room temperature for thirty minutes, the
leukocyte-rich supernatant was aspirated and washed three
times (400 x G, 10 min) with 10 ml of Hank's balanced
salt solution (BSS) (GIBCO, Grand Island, New York)
which contained Ca^{++} and Mg^{++} but no phenol red. Following
washing, the leukocytes were resuspended in BSS at a
concentration of 10^6 PMNL per ml unless otherwise noted.
These preparations contained 40-70% PMNL and 30-60%
mononuclear leukocytes, with a ratio of approximately
10-100 erythrocytes per leukocyte.
 A second method for isolation of mixed leukocytes,
devoid of erythrocytes, was developed when we found that
the conventional techniques resulted in clumping of
leukocytes, and interference with the estimation of the
exact concentration of phacogytes and quantitation of
chemiluminescence. Methyl cellulose (0.5 ml, 2% w/v
in normal saline), was added to 8 ml of heparinized venous
blood to enhance erythrocyte sedimentation. Sedimentation
was allowed to proceed for 1 hr at room temperature. The
leukocyte rich supernatant so obtained was mixed with
2 volumes of warm (37°C) lysing solution (0.155 M NH_4Cl,
10 mM $NaHCO_3$, and 0.3 mM EDTA). This mixture was left to

stand for 5 min at 37°C in a water bath, and then
centrifuged (400 x G) for 10 min. The leukocyte pellet
was gently resuspended in warm lysing solution by drawing
the pellet 5 or 6 times into a Pasteur pipette containing
lysing solution. This mixture was then incubated for a
further 5 min at 37°C, centrifuged (400 x G, 10 min) and
the pellet gently resuspended in warm, sterile, normal
saline. This preparation was centrifuged (300 x G, 10 min)
and the pellet resuspended in saline, centrifuged again
(300 x G, 10 min) and the pellet resuspended in warm
BSS at a concentration of 1.0 x 10^6 PMNL per ml. These
preparations contained 40-70% PMNL and 30-60% mononuclear
leukocytes, of which 60-80% were lymphocytes and the
remainder monocytes.

The third preparation, consisting of 98-100% pure
mononuclear leukocytes was obtained using an adaptation of
the method of Boyum (12). Heparinized venous blood was
diluted with an equal volume of saline and carefully
layered on a "cushion" of Ficoll-Hypaque. The layered
tubes were centrifuged (400 x G, 30 min, 20°C) in a
swinging bucket rotor. The leukocyte layer at the plasma-
Ficoll-Hypaque interface was aspirated, mixed with 5
volumes of BSS, centrifuged (600 x G, 20 min), gently
resuspended in warm BSS (10 ml), centrifuged (400 x G,
20 min), and finally resuspended in warm BSS at a concentration of 1.0 x 10^6 leukocytes per ml. These preparations
contained 98-100% mononuclear leukocytes, 0.2% PMNL,
and no erythrocytes. Of the mononuclear leukocytes,
60-80% were lymphocytic by morphological criteria, and the
remainder were monocytes.

The fourth preparation, consisting of 93-98% PMNL
was obtained using the above methodology, but isolating the
PMNL and erythrocytes which had sedimented to the base of
the Ficoll-Hypaque containing tubes. The PMNL were seen
to form a thin "buffy coat" above the erythrocytes. The
buffy coat was aspirated with a Pasteur pipette, mixed
with 5 volumes of warm lysing solution, and incubated
for 5 min at 37°C. This mixture was centrifuged (400 x G),
the pellet gently resuspended in lysing solution, and
incubated for 5 min at 37°C before being centrifuged
(400 x G) and resuspended in BSS twice more in sequence.
The final concentration was 1.0 x 10^6 PMNL per ml unless
otherwise noted, and the preparations contained 93-98%
PMNL, and 2-7% mononuclear leukocytes with no contaminating
erythrocytes.

Packed Erythrocytes: Heparinized venous blood was refrigerated at 4°C for 72 hours. The sedimented cells were washed three times in BSS (400 x G, 10 min) and pelleted by centrifugation (400 x G, 10 min). The final preparation consisted of approximately 10^6 erythrocytes per mm^3.

Particles and Opsonization: *Candida albicans* was prepared from dry yeast (Standard Brands Ltd., Montreal, Canada) by boiling for 10 min in BSS, filtering through sterile gauze, and suspending in BSS at a concentration of 2 x 10^8 yeast cells/ml. For opsonization, the yeast preparation was diluted to 20% with fresh serum and incubated at 37°C for 30 min with gentle agitation.

Staphylococcus aureus and *Escherichia coli* were prepared from overnight cultures in trypticase soy broth. Bacteria were pelleted by centrifugation, and resuspended in BSS at a concentration of 1.0 x 10^9 organisms/ml (determined by optical density). For opsonization, the bacteria were pelleted by centrifugation (1,000 x G, 10 min), resuspended in fresh serum (0.5 ml per ml pelleted), and incubated for 30 min at 37°C.

Zymosan was purchased from NCI Pharmaceuticals (Plainview, N.Y.) and suspended in BSS at a concentration of 100 mg/ml. For opsonization, 100 µl of the zymosan mixture was mixed with 0.4 ml serum (final concentration - 20 mg zymosan/ml serum), and incubated for 30 min at 37°C.

Chemiluminescence Assays: Background CPM for each vial used in each experiment was determined by monitoring the chemiluminescence of the vial containing 10 ml BSS with the appropriate concentration of opsonized particles. Vials containing the desired concentration of leukocytes in 10 ml BSS were placed into the counter and the leukocyte background chemiluminescence determined. Particles were then added at a concentration of 1 ml of opsonized *C. albicans*, 100 µl of opsonized *E. coli*, *S. aureus*, or zymosan (unless otherwise noted), and readings in CPM were recorded at 0.2 min intervals for at least 1 min, and at minute intervals thereafter for at least 30-60 min. Representative background readings were <10,000 CPM for empty vials, 11,000-13,000 CPM for vials with 10 ml of BSS, 18,000-25,000 CPM for vials with BSS and leukocytes. The background counts of vials with particles in 10 ml BSS were subtracted from background of vials with BSS alone to determine the net particle background. Particle background was added to the background of vials containing leukocytes in BSS to determine total background, and this figure was substracted from counts obtained after addition

of particles to leukocytes (CPM over background). Counts
above total background were plotted as a function of time.

Results

In preliminary experiments, the first leukocyte
preparation, which contained mixed leukocytes and contaminating erythrocytes, was used. Variable results were
found, and it was concluded that erythrocyte contamination
might be a possible source for quenching of photon
emissions. To confirm this, we added increasing amounts
of additional packed erythrocytes to 3 of 4 vials containing
10 ml of 10^6 PMNL per ml with a ratio of approximately
10 erythrocytes/leukocyte. One vial received no additional
packed erythrocytes, and the other three vials received
100, 200, and 300 µl of packed erythrocytes respectively.
Opsonized *C. albicans* was added to each vial and the
chemiluminescence monitored. The results (Fig.2) show
that marked quenching of the chemiluminescence response
occurred with increasing concentrations of erythrocytes.
Thus, it was concluded that leukocyte preparations devoid
of erythrocytes were necessary for precise quantitation
of chemiluminescence. Varying the method for isolating
leukocytes did not alter the ability of the leukocytes
to chemiluminesce (Fig. 3). This was shown by studying
chemiluminescence of leukocytes isolated by the first
method, with those leukocytes separated by the second
method after the addition of opsonized zymosan.

Chemiluminescence Response to Different Particles:
The Chemiluminescence response of leukocytes isolated
by the second method to opsonized and unopsonized *C.
albicans* is illustrated in Figure 3. Opsonization is
required for a maximal chemiluminescence response,
presumably because unopsonized particles are less readily
phagocytozed (13). Phagocytosis is also impaired in
media devoid of Ca^{++} and Mg^{++} (13). In one experiment,
leukocytes were suspended in BSS without Ca^{++} and Mg^{++}
and mixed with a preparation of opsonized *C. albicans,*
in which the cations had been removed, after opsonization,
by addition of 2.5 mg/ml EDTA (disodium ethylenediamine
tetraacetic acid). This mixture did not emit photons
comparable to the control preparation consisting of leukocytes in media containing Ca^{++} and Mg^{++} and opsonized
C. albicans (Fig.4).

The chemiluminescence response of leukocytes isolated
by the second method to opsonized *C. albicans, E. coli,
S. aureus,* and zymosan is illustrated in Figure 5.

Leukocyte Preparations

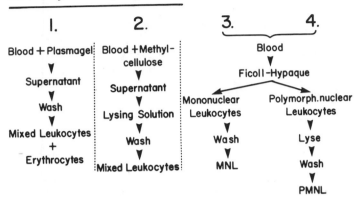

Figure 1.
Methods used to isolate leukocytes from heparinized
venous blood.
MNL - Mononuclear leukocytes

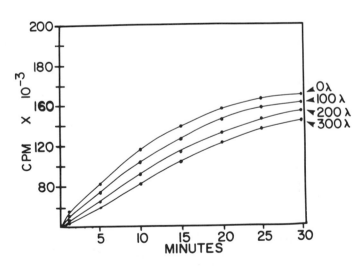

Figure 2.
Color quenching of leukocyte chemiluminescence by erythrocytes.
Chemiluminescence of leukocytes prepared by method 1
(Fig. 1) to opsonized *C. albicans*, as shown by top line,
was markedly reduced by the addition of 100 µl, 200 µl, or
300 µl of packed erythrocytes to leukocyte suspensions.

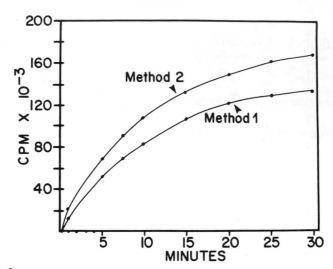

Figure 3.
Chemiluminescence response of leukocytes to opsonized zymosan.
Leukocytes were prepared from the same blood sample depicted in Figure 2, by method 1, containing contaminating erythrocytes, or by method 2, free from erythrocytes (Fig. 1).

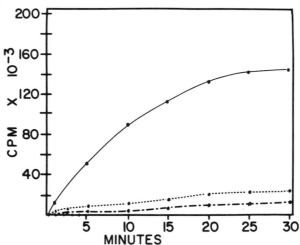

Figure 4.
Inhibition of phagocyte chemiluminescence response to *C. albicans* (——) by use of media devoid of cations (–·–·–·–), or by use of unopsonized particles (-----). In the latter two cases, microscopic examination of the incubation mixture revealed markedly decreased phagocytosis of particles by phagocytes.

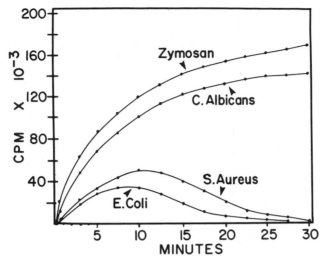

Figure 5.
Chemiluminescence response of phagocytes elicited by
opsonized zymosan, *C. albicans*, *S. aureus*, or *E. Coli*.
Leukocytes were prepared as described in method 2 (Fig. 1).

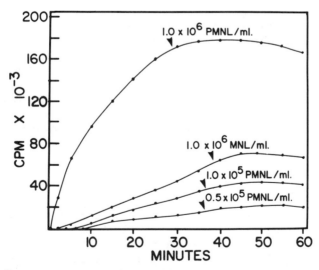

Figure 6.
Chemiluminescence response by PMNL and mononuclear leukocytes isolated by methods 3 and 4 (Fig. 1). Leukocytes
were suspended in BSS at the concentrations indicated,
and tested using opsonized zymosan.

The response to bacteria was short-lived, whereas the response to *C. albicans* and zymosan lasted for up to 30 min and then fell gradually. A possible explanation for this might be that the larger yeast particles are subject to a continual attack by antimicrobial systems, whereas the smaller bacteria are phagocytozed, killed and digested more quickly.

Monocyte Chemiluminescence: The mechanism for bacterial killing by human monocytes might be similar to that of PMNL (14). However, no published information is available concerning the chemiluminescence of monocytes. In our hands, mononuclear leukocytes, isolated by the third method, containing <3% PMNL contamination, emitted photons in response to *E. coli* and zymosan. In an attempt to ensure that contaminating PMNL did not cause this response, chemiluminescence response of suspensions of PMNL (1.0×10^5 and 0.5×10^5 per ml) isolated by the fourth method, were studied, after the addition of opsonized zymosan (Fig. 6). A PMNL suspension of 1.0×10^5 would be equivalent to a 10% PMNL contamination of the monocyte suspension, while a suspension of 0.5×10^5 PMNL would be equivalent to a 5% PMNL contamination. Neither of these concentrations of PMNL gave sufficient photon emission to account for the observed monocyte response. The response of 1.0×10^6 PMNL/ml isolated from blood of the same individual as the above mononuclear leukocyte suspension is also shown in Figure 6 for comparison. PMNL responded more quickly and with greater intensity than the mononuclear leukocytes. However, the majority of the mononuclear leukocytes in the suspension were nonphagocytic lymphocytes, and therefore the response in terms of CPM per cell cannot be directly compared.

Variability: The day-to-day, and individual-to-individual, variability of chemiluminescence response of PMNL and monocytes was studied. The chemiluminescence response to opsonized zymosan of PMNL and mononuclear leukocytes, isolated from the blood of five different healthy individuals and assayed on one day, is shown in Figures 7 and 8. The response of mononuclear leukocytes showed little variability for the first 30 min, but the response of PMNL showed much greater variability. However, it was not thought that this variability in PMNL response would preclude the use of chemiluminescence for quantitative comparisons in a clinical setting. The variability of the response to opsonized zymosan by PMNL and mononuclear leukocytes isolated from the blood of one

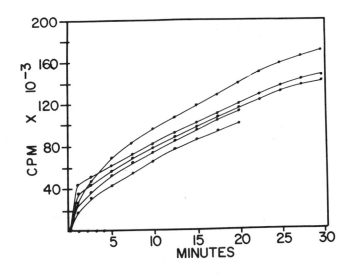

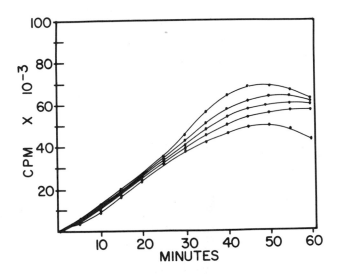

Figures 7 and 8.
Variability in the chemiluminescence response by PMNL
(Fig. 7) and mononuclear leukocytes (Fig. 8) to opsonized
zymosan. Leukocytes were isolated by methods 3 and 4
(Fig. 1) from the blood of five healthy individuals, and
tested on the same day. Note the difference in axis scale
of Figure 6 as compared to Figure 7.

healthy individual on three consecutive days was minimal (Figs. 9 and 10).

Discussion

Chemiluminescence of human PMNL has previously been reported (17). However, few studies have commented upon the importance of the quenching effect of contaminating erythrocytes on chemiluminescence, and none, to our knowledge, have considered the contribution of contaminating mononuclear leukocytes to PMNL chemiluminescence. The studies reported here demonstrate that precise and consistent chemiluminescence quantitation is difficult, if not impossible, if heterogeneous populations comprising PMNL, mononuclear leukocytes and contaminating erythrocytes of varying numbers are used as samples. As reported here, methods are available for the isolation of relatively homogeneous suspensions of PMNL and mononuclear leukocytes, which are virtually devoid of contaminating erythrocytes. Such homogeneous suspensions are particularly useful for studying chemiluminescence of phagocytic cells, and should have clear-cut clinical application - particularly, if phagocyte chemiluminescence proves to be a measurement of the rate and total microbicidal capacity of phagocytes. In addition, the chemiluminescence technique, using the methods described herein, represents an important advance in terms of accuracy, ease, and consistency when compared to conventional quantitative bactericidal assays.

Stossel (13) indicates criteria which must be met in order to assert that a particular assay measures phagocytosis and intracellular killing. These include:
(1) Inhibition of the reaction at ice bath temperatures
(2) Demonstration that the rate of particle ingestion decreases as particle load increases
(3) Demonstration that no reaction occurs at zero time at physiological temperatures.
Presently, we are performing experiments to meet these criteria, as well as to determine whether chemiluminescence parallels bactericidal activity as assayed by conventional methods. In addition, we are also investigating the effect of various metabolic and bactericidal inhibitors on the chemiluminescence response of human phagocytes in an attempt to delineate the biochemical nature of the chemiluminescence assay.

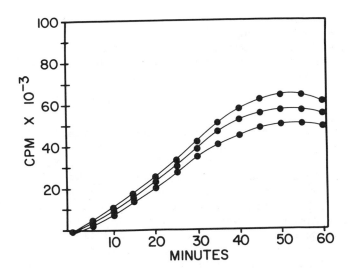

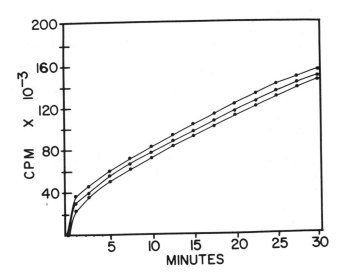

Figures 9 and 10.
Variability in the chemiluminescence response by PMNL
(Fig. 9) and mononuclear leukocytes (Fig. 10) to opsonized
zymosan. Leukocytes were isolated by methods 3 and 4
(Fig. 1) from the blood of one healthy individual on
three consecutive days.

Acknowledgements
    This work was supported in part by the Research
Committee of the Provincial Cancer Hospitals Board,
Alberta, the Dr. Mervin Laskin Fellowship in Cance (DE),
and by the MRC of Canada (Grant MA-5904).

References
1.   R.C. Allen, R.L. Stjernholm and R.H. Steele, Biochem.
     Biophys. Res. Commun. 47, 679 (1972).
2.   R.B. Johnston, B.B. Keele, H.P. Misra, L.S. Webb,
     J.E. Lehmeyer and K.V. Rajagopalan in The Phagocytic
     Cell and Host Resistance, p. 61, (J.A. Bellanti
     and D.H. Dayton, Eds.). New York: Raven Press (1975).
3.   S.J. Klebanoff, Semin. Hematol. XII, 117 (1975).
4.   P.G. Quie, Semin, Hematol. XII, 143 (1975).
5.   B.D. Cheson, R. Christiansen, R. Sperling, B. Kohler
     and B.M. Babior, Blood 46, 1016 (1975) (Abstract).
6.   E.A. Ogryzlo in Photophysiology, Vol. V, p. 35
     (C. Giese, Ed.). N.Y.: Academic Press (1970).
7.   R.C. Allen and R.H. Steele, Fed. Proc. 32, 478 (1973).
8.   C.S. Foote, R.W. Denny and L. Weaver, Ann. N.Y.
     Acad. Sci. 171, 139 (1971).
9.   S.J. Klebanoff in The Phagocytic Cell and Host
     Resistance, p. 45 (J.A. Bellanti and D.H. Dayton, Eds.)
     New York: Raven Press (1975).
10. J.M. McCord, B.B. Keele and I. Fridovich, Proc. Natl.
     Acad. Sci. 68, 1024 (1971).
11. L.R. DeChatelet, C.E. McCall, L.C. McPhail and
     R.B. Johnston, J. Clin. Invest. 53, 1197 (1974).
12. A. Boyum, Tissue Antigens 4, 269 (1974).
13. T.P. Stossel, Semin, Hematol XII, 83 (1975).
14. M.J. Cline in The White Cell, P. 493. Cambridge:
     Harvard Univ. Press (1975).

ON THE USE OF LIQUID SCINTILLATION COUNTERS
FOR CHEMILUMINESCENCE ASSAYS IN BIOCHEMISTRY

E. Schram, F. Demuylder, J. De Rycker and H. Roosens
Vrije Universiteit Brussel, Laboratorium voor Biochemie
Brussels, Belgium

ABSTRACT

In order to help interpreting the results obtained
with commercial firefly luciferase preparations, their active components were assayed and the conversion of ATP studied
with labelled nucleotides. A quenched scintillating solution
containing tritium was used as a reference for photon counting. The use of scintillation counters for the assay and
study of oxygenases and oxidases is discussed.

INTRODUCTION

About ten years have elapsed since the use of liquid scintillation counters was extended to the measurement
of bioluminescence in the microassay of ATP with firefly
luciferase. Although such instruments may have superfluous
features they have ever since proven reliable and adequate
not only for the assay of ATP and related substances with
firefly luciferase, but also for that of flavine and pyridine
nucleotides with the bacterial system. Several survey articles have covered the status of the art in recent years
(1, 2, 3). In the present paper we wish to report more
specifically on the use of scintillation counters for the
study and assay of other biochemical reactions, involving
oxidases and oxygenases, as well as on some recent experience with luciferases.

LUMINESCENCE MEASUREMENTS

A problem often encountered in luminescence coun-
tings is the lack of a stable source delivering single discrete photons. Such a source is essential when checking for
the efficiency or reproducibility of scintillation counters
and when optimizing the settings for high voltage, amplifi-

243

cation and discriminators. As mentioned earlier (2, 3), one
cannot rely on the settings found adequate for the counting
of tritium. The use of luminol as a standard has been described by Lee et al.(4). However, the concentrations used by
these authors produce light levels much higher than those
suitable for scintillation counters. Scaling down the reaction
is not an easy matter because of the reagent background (5)
and of the tricky character of the reaction. A stable luminescence source, based on the use of a carbon-14 source in a
liquid scintillator, was developed by Hastings et al. (6,7).
This kind of standard is quite adequate in conjunction with
analog instruments but is not suitable as a refernce source
for photon counting because of the production of multi-photon
flashes. The use of a light-emitting diode driven by a small
mercury battery has been suggested by Stanley (3). When trying
to solve the problem ourselves, satisfactory results were obtained by incorporating a tritium source in a liquid scintillator, quenched so as to emit only single photons. Adequate
quenching is easily achieved by the addition of carbon tetra-
chloride until the number of counts registered with the coincidence switched on falls to 1%, or preferably less, of that
registered with the coincidence left out. The number of two-
photon pulses falling on either one of the photomultipliers
is not exactly the same as the number of such pulses distri-
buting themselves between the two photomultipliers (8). The
number of coincidence pulses is therefore not perfectly representative but offers a sufficient approximation. A rather
high degree of quenching is necessary to eliminate two-photon
pulses (> 99 %) and the initial radioactivity should therefore be chosen in accordance. Care should also be taken to
repeat the test when changing the high voltage or discrimination levels as the ratio of two-photon vs. one-photon pulses
may increase with higher voltages or higher discrimination
levels. The validity of the present method was checked by
analyzing the energy spectrum of the emitted light. The set-
up used for this purpose was equipped with an RCA 8850 photomultiplier, specially designed for single-photon counting,
and a SA41 400-channel analyzer. The quenched tritium spectrum was compared with the phosphorescence spectrum of a
scintillation vial cap that had been exposed to intense light
(see figure 1). This spectrum also matches that obtained
for firefly bioluminescence. If desirable, the emission spectrum of the present reference solution could be made to match
more closely that of the luminescent reaction being studied
by the addition of a suitable secondary scintillator.

Any tritium-labelled compound available in the laboratory is in principle eligible for the preparation of a refe-

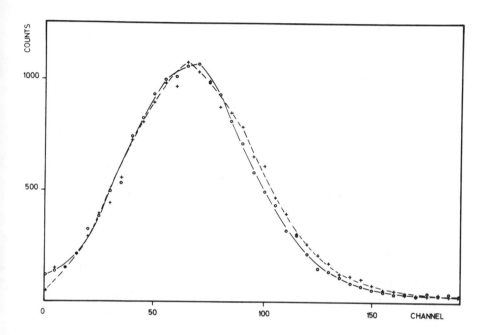

Fig. 1. Energy spectrum recorded with SA41 multichannel
analyzer:
   a) tritium in scintillating solution containing
      PPO and POPOP and quenched with CCl4 (———)
   b) phosphorescence of scintillation vial cap (----)

rence luminescence source. We wish however to stress the fact
that when hydrophilic substances are dissolved in scintillating solutions containing emulsifying agents as Triton-X-100,
temperature has to be controlled very carefully. Several authors have indeed shown the counting efficiency of emulsions
to be rather sensitive to slight temperature changes (9).
This sensitivity is still more apparent when single photons
are counted with the coincidence left out, amounting to a
3 % increase of the count rate for a one centigrade drop. A
lower degree of temperature dependence was observed when
using tritium-labelled toluene in regular scintillating solutions.

Our luminescence standard allowed us to check our
counting equipment for long-term stability as well as for
short-lived fluctuations. For a Packard model 2002 counter
the fluctuations observed over one minute periods were somewhat larger than statistically predicted (half of them exceed twice the standard deviation) but are not in general
higher than 1 %. The absolute calibration of the standard
solution just described would provide an easy tool for the
estimation of the quantum efficiency of bio- and chemiluminescent reactions, and which can be used in precisely the
same geometrical conditions as the sample being assayed.

LUMINESCENCE PRODUCED BY LUCIFERASES

The use of scintillation counters for biolumines-
ence measurements in analytical biochemistry has been discussed by Stanley (see present volume) and we shall therefore
restrict ourselves to a few recent contributions from our
laboratory.

First of all we wish to mention the work performed by E. Gerlo and J. Charlier (10) on the separation of the
enzymes involved in bacterial bioluminescence. The MAV strain
used for this purpose, kindly supplied by J.W. Hastings and
at present identified as Beneckea Harveyi, was preferred to
the wild type Photobacterium Fischeri because the luciferase
is more easily separated from the FMN reductase (NADH-dehy-
drogenase). In the course of this separation it occurred to
Gerlo and Charlier that two FMN reductases were actually present in the extract, specific for NADH and NADPH. These enzymes were separated by chromatography on DEAE-Sephadex and
purified by a factor of 90 and 140. They were further identified by studying several of their characteristics: molecular
weight, thermostability, relative specificity for NADH and
NADPH, dissociation constants of the binary and ternary com-
plexes (with FMN). These findings open new possibilities

for the selective assay of NADH and NADPH, which can be individually measured in the presence of a 50-fold or even higher
excess of the other nucleotide.

As far as <u>firefly</u> luciferase is concerned it should
be observed that most routine assays are still performed with
commercial enzymes that are impure. The practical implicati-
ons of this on the specificity and the time-course of the luminescence have been reexamined very thoroughly in a recent
paper by Lundin and Thore (11). In our opinion interpretation
of the results is often impeded by the fact that the enzyme
content of luciferase preparations is not known. They may
further contain varying amounts of luciferine. In order to
characterize commercial preparations in this respect a simple
method was used for the rapid chromatographic fractionation
and the fluorimetric titration of the enzyme with dehydroluciferine. One ml samples, containing the equivalent of ca.
10 mg firefly organs, are brought on a 1.28 x 10 cm Sephadex
G-25 column (particle-size 50-150 μ) and eluted successive-
ly with glycylglycine buffer, pH 7.4, 0.025 M, and pure water
to recover the luciferine (the present technique was adapted
from Nielsen and Rasmussen, 12). Fractions 1.5 ml in size
are analyzed for their fluorescence and, after addition of
ATP, Mg^{++} and luciferine, for their bioluminescence (see
figure 2). The first peak shows a spectrum typical for proteins and contains all of the luciferase together with contaminating enzymes. The second peak, identified by its strong
fluorescence at 414 nm is well separated but was not identified thus far. The fractions containing luciferase are pooled
and aliquots titrated with dehydroluciferine in an Aminco
spectrofluorimeter, as described by DeLuca and McElroy (13).
Dehydroluciferine was prepared by refluxing 10 mg of luciferine in one ml of 1 N sodium hydroxide for 4 hours (14) and
isolated by chromatography on Sephadex G-25 under conditions
similar to those described above.

The firefly extracts tested with the present method were purchased from Sigma (cat.nr. FLE-50) and from
Boehringer (cat.nr. 15480). Assays performed on the luciferase fraction gave similar figures for the enzyme contents
of both preparations, i.e. 3.8 and 2.7 x 10^{-7} equivalents per
100 mg organs. The figures were also similar when expressed
as a function of the protein content, based on the optical
density at 280 nm. Luciferine was estimated by the spectrophotometric assay of the pooled fractions obtained upon elu-
tion with water. The fluorescence excitation and emission
spectra were also recorded and found typical for luciferine.
Higher figures were obtained for Sigma than for Boehringer
relative to the enzyme content (the molar ratios were 1.0

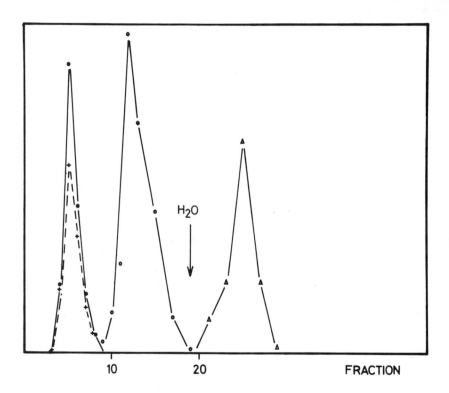

Fig. 2. Chromatography of commercial firefly luciferase.
Ordinates (arbitrary units):
Fluorescence: (●) excit. 280 nm em. 350 nm
           (o)   "   348   " 414
           (△)   "   330   " 530
Bioluminescence: (+)

and 0.2 respectively). It should on the other hand be mentioned that the Boehringer preparation is perfectly soluble
while that of Sigma contains suspended material which sticks
on top of the column, or sediments upon centrifugation. Part
of the luminescence was found to be associated with this precipitate. Due to the limited number of batches at hand no
systematic tests could be performed up to now and the present
results may therefore not be statistically valid. From the
figures obtained it may nevertheless be concluded that the
amount of enzyme routinely used for the assay of ATP with
scintillation counters is several orders of magnitude higher
than that of ATP. End-product inhibition is therefore not
likely to occur under the usual conditions of assay and is
at least not responsible for the gradual decrease of the luminescence. The complex time-course of the luminescence cannot be accounted for either, on the mere basis of the ex-
haustion of the substrate. Experiments performed with ^{14}C-
labelled adenine nucleotides on the enzyme fraction obtained
by our procedure have indeed shown the radioactivity of ATP
to appear rather rapidly in ADP. The observed facts are in
favour of an equilibrium between the several adenine nucleotides, which might account for the rapid initial decrease and
the long tailing of the luminescence curve obtained with un-
purified luciferase preparations.

CHEMILUMINESCENCE INITIATED BY OXIDASES AND OXYGENASES

In the past years study of chemiluminescence in
biochemistry has often concentrated on a series of reactions
catalyzed by specific "luciferases". Beside the fact that
these reactions show fascinating biological aspects, their
specificity and high quantum yield make them suitable for a
wide array of analytical applications, for which liquid scintillation counters are now routinely used. Bioluminescence
is, however, not restricted to these specific cases and low
level luminescence has been observed for many years in living
cells not suspected of containing any luciferase. Light emission has been reported in mitochondrial preparations, in mi-
crosomal extracts, in the course of phagocytosis (see
English et al., present volume), aso. Although oxidation
reactions have long been considered responsible for these
phenomena, interest in these reactions has recently increased, due to several important discoveries:
- the involvement of superoxide ions (O_2^-) in several biochemical reactions and physiological processes;
- the ubiquitous presence of superoxide dismutase (SOD) in
the cells of aerobic organisms (15, 16).

249

New insights are developing as far as the direct involvement
of oxygen in biological oxidations is concerned and the clas-
sification of some of the responsible enzymes might even have
to be reconsidered. A comprehensive and critical survey of
these enzymes is to be found in a publication of Hayaishi(17).
        The reduction of O_2 concomitant with the oxida-
tion of a substrate molecule can be considered to occur in
discrete steps:

$$O_2 \longrightarrow O_2H \longrightarrow H_2O_2 \longrightarrow H_2O$$

If O_2 is not fully reduced to water, as occurs with cytochro-
me oxidase, O_2 and H_2O_2 are likely to appear as intermediates.
Superoxide radicals are rather unstable and will disproportio-
nate spontaneously according to the following reaction: (18)

$$O_2^- + O_2^- + 2\ H^+ \longrightarrow H_2O_2 + O_2$$

The speed of this reaction is, however, much dependent on the
dissociation of O_2H (pK = 4.8): (19)

dismutation of O_2H rate constant = 7.6×10^5 $M^{-1}.sec^{-1}$

     "   " O_2^- "   " < 100 $M^{-1}.sec^{-1}$

These facts account for the existence of superoxide ions at
physiological pH and probably also for the toxicity of oxygen.
The presence of superoxide dismutase, which catalyzes the
dismutation of these ions, is therefore essential for the sur-
vival of aerobic cells. The study of reactions involving superoxide ions may be complicated by their interconversion in
the presence of H_2O_2 (produced or not by the dismutation of
superoxide radicals) according to the Haber-Weiss mechanism (20):

$$O_2^- + H_2O_2 + H^+ \longrightarrow O_2 + H_2O + OH^\cdot$$
$$OH^\cdot + H_2O_2 \longrightarrow H_2O + O_2^- + H^+$$

Hydroxyl radicals are potent oxidants and might therefore be
the molecular species ultimately responsible for the toxicity of oxygen.
        Superoxide radicals can trigger the production of
light in several ways. First of all their dismutation will
produce oxygen in the singlet excited states $^1\Delta_g$ or $^1\Sigma_g^+$.
The energy of transition to triplet oxygen in the ground state corresponds to light emission in the red region (760 or
1260 nm) which is normally not registered by regular photomultipliers. Emission may however be detected through the formation of excited dimers (oxygen excimers). It has further
been observed that the luminescence due to superoxide ions
could be enhanced by bicarbonate anions. This effect has been

ascribed by Stauff (18) to the recombination of intermediary carbonate radicals. Much higher amplification can still be obtained by the addition of luminol (LH_2) which, according to present knowledge would react in the following way (21, 22):

$$LH_2 \longrightarrow LH^- + H^+ \quad (\text{ in basic medium })$$

$$LH^- + OH^\cdot \longrightarrow LH^\cdot + OH^-$$

$$LH^\cdot + O_2^{\cdot-} \longrightarrow LO_2H^- \longrightarrow N_2 + P^x$$

$$P^x \longrightarrow P + h\nu$$

$$LH^- + O_2^{\cdot-} + 2 H^+ \longrightarrow LH^\cdot + H_2O_2$$

$$LH^- + H_2O_2 \longrightarrow LH^\cdot + OH^- + OH^\cdot$$

$$(\ P = \text{aminophthalate ion} \ )$$

As may be deduced from the above equations and from the Haber-Weiss reaction, superoxide radicals will be able to generate luminol luminescence by themselves, contrary to H_2O_2. This last substance will, however, enhance the luminescence when superoxide ions are already present. All these facts should be carefully born in mind when using luminol for the study of the luminescence induced by oxidases and oxygenases.

Superoxide radicals are produced in a variety of biochemical systems (xanthine-oxidase, aldehyde-oxidase, flavoprotein oxygenases, aso) while other reactions will preferentially generate hydrogen peroxide (D-amino-acid oxidases, glucose oxidase aso.). The luminescence associated with these reactions or which can be induced by the addition of luminol, has become a very sensitive tool, not only for analytical purposes, but also for the study of the basic mechanisms involved in these reactions. In the last instance the appearance of specific oxygen intermediates can often be inferred from the influence of adequate inhibitors on the time-course of the luminescence. Examples of such inhibitors are SOD (superoxide dismutase) for superoxide radicals, catalase for hydrogen peroxide, scavengers like ethanol for hydroxyl radicals also. As for the reactions involving luciferases, scintillation counters have proven very useful for the kinetic study of such biochemical systems. Because of the interconversion of the oxygen intermediates the results should nevertheless be interpreted very cautiously and the answer will not always be straightforward.

The systems tested in our laboratory involve xanthine-oxidase, horse-radish peroxidase and reduced FMN. The xanthine-oxidase system is well known for the production

251

of superoxide radicals and its luminescence has been studied
by means of scintillation counters by Arneson (23) and by
Schaap et al. (24). While our own experiments with luminol
and inhibitors were in progress similar work was reported by
Hodgson and Fridovich (22)showing the inhibitory effect of
both catalase and superoxide dismutase on the luminescence.
We could further observe a cooperative effect between these
inhibitors as well as the inhibitory effect of ethanol. Spectrophotometric recording of the urate formation shows that
- the formation of superoxide radicals is a result of the
  enzymatic reaction and inhibition of the luminescence does
  not interfere with the actual oxidation of xanthine;
- the luminescence produced can be turned to use as a very
  sensitive method of assay for xanthine or xanthine-oxidase.
In all these experiments the enzymatic reaction is performed
at pH 10.2 in order to fit with the pH requirements of the
luminol reaction.

   The reaction products of the xanthine-xanthine oxidase system (uric acid and hydrogen peroxide) happen to be
substrates for horse-radish peroxidase. When mixing both enzymes a coupled system was obtained which produced a luminescence increased by a factor of 5-10 compared with that obtained
for xanthine-oxidase alone.

   Reduced FMN is very unstable and reoxidizes spontaneously with a half-life of the order of a second. Because of
the known production of hydrogen peroxide in this reaction it
seemed interesting to study the effect of inhibitors on the
luminescence produced by the addition of luminol (25). FMN
was reduced with dithionite according to Gerlo and Charlier
(10). Inhibition was observed for both superoxide dismutase
and catalase.

   All our luminescnce measurements were performed
with either a Packard (model 2002) or an Intertechnique
(model SL20) scintillation counter.

   The use of oxygenase induced chemiluminescence for
analytical purposes is best illustrated by the method described by Bostick and Hercules (26) for the automated determination of glucose with immobilized glucose oxidase. Hydrogen
peroxide, but no superoxide radicals, is produced in this reaction and potassium ferricyanide has therefore to be added
to the luminescent system. Another feature of this method is
the fact that the oxidation of glucose and the luminescent
reaction are occurring sequentially with the intercalation of
a pH jump. The method would seem to be applicable to other
substrates yielding hydrogen peroxide in neutral solution.

REFERENCES
1. H.H. Seliger in Chemiluminescence and Bioluminescence p. 461 (M.J. Cormier, D.M. Hercules and J. Lee, Eds.) New York and London: Plenum Press 1973
2. E. Schram in Liquid Scintillation Counting p. 383 (P.E. Stanley and B.A. Scoggins, Eds.), New York and London: Academic Press 1974
3. P.E. Stanley in Liquid Scintillation Counting Vol. 3 p.253 (M.A. Crook and P. Johnson, Eds.) London: Heyden 1974
4. J. Lee, A.S. Wesley, J.F. Ferguson and H.H. Seliger in Bioluminescence in Progress p.35 (F.H. Johnson and Y. Haneda, Eds.) Princeton: Princeton Univ. Press 1966
5. H.H. Seliger, Photochem. and Photobiol. $\underline{21}$, 355 (1975)
6. J.W. Hastings and G. Weber, J.Opt.Soc.Am. $\underline{53}$, 1410 (1963)
7. J.W. Hastings and G.T. Reynolds in Bioluminescence in Progress p.45 (F.H. Johnson and Y. Haneda, Eds.) Princeton: Princeton Univ. Press 1966
8. P.J. Malcolm and P.E. Stanley in Liquid Scintillation Counting p.77 (P.E. Stanley an B.A. Scoggins, Eds.) New York and London: Academic Press 1974
9. R.H. Benson, Analyt. Chem. $\underline{38}$, 1353 (1966)
10. E. Gerlo and J. Charlier, Eur. J. Biochem. $\underline{57}$, 461 (1975)
11. A. Lundin and A. Thore, Analyt. Biochem. $\underline{66}$, 47 (1975)
12. R. Nielsen and H. Rasmussen, Acta Chem. Scand. $\underline{22}$, 1757 (1968)
13. M. DeLuca and W.D. McElroy, Biochemistry $\underline{13}$, 921 (1974)
14. E.H. White, F. McCapra and G.F. Field, J. Am. Chem. Soc. $\underline{85}$, 337 (1963)
15. J.M. McCord and I. Fridovich, J. Biol. Chem. $\underline{244}$, 6049 (1969)
16. I. Fridovich, Ann. Rev. Biochem. $\underline{44}$, 147 (1975)
17. O. Hayaishi in Molecular Mechanisms of Oxygen Activation p. 1 (O. Hayaishi Ed.) New York and London: Academic Press 1974
18. J. Stauff, U. Sander and W. Jaeschke in Chemiluminescence and Bioluminescence p. 131 (M.J. Cormier, D.M. Hercules and J. Lee, Eds.) New York and London: Plenum Press 1973
19. D. Behar, G. Czapski, J. Rabani, L.M. Dorfman and H.A. Schwarz, J. Phys. Chem. $\underline{74}$, 3209 (1970)
20. F. Haber and J. Weiss, Proc. Roy. Soc. (London) $\underline{A147}$, 332 (1934)
21. E.K. Hodgson and I. Fridovich, Photochem. and Photobiol. $\underline{18}$, 451 (1973)
22. id., Biochem. Biophys. Acta $\underline{430}$, 182 (1976)
23. R.M. Arneson, Arch. Biochem. Biophys. $\underline{136}$, 352 (1970)
24. A.P. Schaap, K. Goda and T. Kimura in Excited States of Biological Molecules p.79 (J.B. Birks, Ed.) London:

Wiley 1976

25. A.M. Michelson, Biochimie 55, 465 (1973)
26. D.T. Bostick and D.M. Hercules, Anal. Chem. 47, 447 (1975)

Acknowledgements

This work was supported by the Belgian Staatssecretariaat voor Wetenschapsbeleid.

INSTRUMENTATION WORKSHOP

PANEL MEMBERS

Mr. B.H. Laney   - Chairman

Dr. J.E. Noakes - Low Background Liquid Scintillation
                    Counters

Dr. P.E. Stanley - Data Processing in Liquid Scintillation
                     Counting

INSTRUMENTATION WORKSHOP: INTRODUCTION

B.H. Laney
Searle Analytic, Inc.
Des Plaines, Illinois 60018

Although the invention of the vacuum triode amplifier
and multiplier phototube were key ingredients to achieve
the sensitivity necessary for technical feasibility, it was
the application of the coincidence technique in 1950 by
Reynolds which made liquid scintillation spectrometers (LSS)
commercially practicable. Recognition of tritium as a
valuable biological tracer and the automation of LSS's in
1957 created a preferential demand for LSC's over ion
chamber and 2π gas flow proportional counters. Invention
of the external standard by Fleishman and Glazunov in the
U.S.S.R. in 1961 and its subsequent automation by ANSITRON
in 1964 was a major advance in simplifying the determination
of efficiencies, generally replacing the internal standardization method.

During the period from 1954 to 1968, major improvements
in detection efficiency occurred principally due to the
development of multiplier phototubes (MPT) with higher
quantum efficiency and lower background. Graded shielding
and anti-coincidence guard detectors have also been used to
reduce background rates to several counts per minute.

Developments during the last decade have been directed
toward improving the accuracy, speed, reliability, cost,
effectiveness and operator convenience of the instruments.
Tray changing mechanisms have been developed to provide for
easier sample loading and more compact instruments. Multi-
user batch operation is available from most manufacturers.
Both pre-programmed and user programmable data reduction
capabilities are also available. Automatic-calibration,
color restoration, and improved quench correction methods
have been developed to reduce potential quench-correction
errors. Future developments will most likely be in the
area of improved data processing and sample handling.

LOW BACKGROUND LIQUID SCINTILLATION COUNTERS

by

John E. Noakes
Geochronology Laboratory
University of Georgia, Athens
Athens, Georgia U.S.A.

Abstract
    Counting low level radioactive samples with a liquid
scintillation counter requires optimized calibration for
low background response, high counting efficiency and
maximum counter stability. Instrument background count rate
can be reduced by almost an order of magnitude without loss
in counting efficiency through the use of special sample
counting vials and selective choice of minimum sample-
cocktail volume. Counter stability can be enhanced by
operating the counter in a balance point mode of counting.
Electronic background noise can be diminished through the
use of faster coincidence timing and selection of matched
low dark current noise photomultiplier tubes. Background
caused from external radiation can be lowered through the
use of an anticoincidence guard and additional massive
shielding. The purpose of this paper is to discuss these
means and the results of their application to optimizing
liquid scintillation counters for low level counting.

    ... Dr. Noakes presented similar information to that he
gave on low background counters at the symposium on liquid
scintillation counting held in Bath, England in 1975. The
proceedings of this symposium which will include Dr. Noake's
presentation are to be published in 1976 by Heyden*. The
reader is directed to these proceedings for a fuller account
of Dr. Noake's work in this field.

*Crook, M.A. and Johnson, P. Liquid Scintillation Counting
volume IV, Heyden, London.

AUTOMATIC DATA PROCESSING IN SCINTILLATION COUNTING

Philip E. Stanley

*Department of Clinical Pharmacology
The Queen Elizabeth Hospital
Woodville, South Australia 5011*

WORKSHOP SESSION

I shall endeavour to present an overview of this topic
and keep in mind the worker who is considering the use of
automatic data processing to assist in handling his data.
I do not intend to dwell on programming or statistics but
rather to present the advantages and pitfalls of the
technique.

In scintillation spectrometry, data processing comes
into two distinct categories. Firstly there is the simple
processing such as is used for calculating counts per
minute, channels ratio or percent free for radioimmunoassays.
Secondly there is the more sophisticated approach which
involves quench correction curves, statistical appraisal
of data and checking for the performance of the sample and
the instrument.

Consider the first of these two categories. It is the
requirement which is in most demand in those laboratories
using scintillation spectrometers. We all know that the
instrument producing data made up of raw counts usually
gets set to count for one or ten minutes with the preset
count terminator adjusted to its maximum. It is thus very
easy for the worker to obtain CPM since no manual division
is required. This is clearly very wasteful in terms of
counter time when most people should be considering a preset
count of say 10^4 as their major terminator. They are of
course resistant to this approach since the data then needs
to be reduced manually. The use of a small dedicated
calculator (on-line) is clearly an advantage in this case
and with the cost of small processors and pocket calculators

getting even lower it seems that these could be offered at a much more attractive price than say five years ago. Such units offer advantages not only for those laboratories where a few samples are counted at a time (say less than 20) but also in those laboratories with a service commitment, say in a hospital where hundreds per day must be processed.

In the second category we consider data processing in its fullest meaning. It is much more involved than the approach just described and should not be embarked upon without considerable forthought. This approach inevitably means the use of some kind of computer which can be programmed by the worker or someone on his behalf. The dedicated on-line computer is what most people think they need to solve their processing problems. I suggest that this is not necessarily the case unless they plan to have several counters on-line at the same time each with a multitude of samples or with a particularly intricate processing problem to carry out. Since counters generally produce only a relatively few numbers every minute and the computer can process these in a fraction of a second it is clear that for 99% of its time the computer has little or nothing to do and it is obviously an inefficient use of the instrument.

The batch processing of data is probably the best approach for most people. Here the data is punched on paper tape and is then accumulated until it is processed at a computer installation. Punched paper tape still seems to be the method of choice as a medium for data transfer. Cassettes and cartridges of magnetic tape are not yet in regular use although they do seem to offer considerable advantages. The lack of availability of a suitable reader at the computer installation could certainly be a drawback and it is of course almost mandatory that a printed copy of the data be made at the same time as it is recorded on the tape. It may be the cost of such a system which precludes its common introduction.

Interactive processing is an excellent way to deal with scintillation data. Here the user can interact with the program and make decisions as to the best mode of action. It is too, if the program is well written, the ideal way for the beginner to become familiar with the program and its do's and its dont's. There is also the advantage just as in the batch system that it can be used for a unique run just as well as for a production run.

Turnaround time is often an item which must be considered against the costs involved. If the worker has access to a large institutional computer installation, the turnaround time may be as long as a couple of days. For some purposes this may be entirely adequate while for others such as in the clinical chemistry laboratory this is quite unsatisfactory. In the latter case the user should perhaps consider the purchase of a small computer. The costs of these is decreasing almost daily and since the unit can be used for processing data from other laboratory instruments this mode of action may be the one of choice. Certainly all that is needed is say 16 or 32 K of core, a high speed paper tape reader and a printer together with a visual display unit. Sometimes it may be advantageous for several groups of people in one institution to make a joint purchase, however the politics of such a decision may have overriding consideration. If a joint purchase is considered it is important to have responsibilities, both financial and departmental, established well in advance since informal arrangements have a habit of being "misunderstood" at a later date.

Let us now turn our attention to some of the features which can be written into a program. These include selection of working parameters e.g. one or two isotopes; which of perhaps several spectrometers has been used to obtain the data; how many standards and what is their DPM; is decay during the counting process significant; how many unknowns; how is efficiency to be determined, by channels ratio or external standardization; In the latter case, the program may have code to decide whether or not the technique is valid, for example external standardization should not be used for heterogeneous samples and channels ratio is not very adequate for measuring the efficiency of tritium.

Next comes the procedure of fitting the curve for the standards so to obtain, for example, an external standard ratio vs efficiency relationship. This seemingly easy task has caused many frustrating problems for the uninitiated user basically because it is not always prudent to obtain a curve which fits all the points very closely. Consider the situation where one standard point has a gross error. Should the curve be constrained to go through that point or should it merely pass close to it and be weighted by all the

263

other points. It would appear that a polynomial expansion, say a cubic, will generally be adequate for many purposes. Carroll and Houser (1) have discussed the problem of curve fitting in an excellent article and some statistical criteria which should be applied in such decision making.

Some workers require to know the standard deviation of their result and this in my opinion is hard to obtain. Radioactive events follow Poisson statistics. However, background counts generally do not and thus there is a statistical dilemma where low activity samples are concerned. Next we must consider vials, their variation in background, their geometry and also the transmission of light through their walls. While these individual features vary only a little they can and do accumulate and of course effect not only the standards but also the unknowns. Then we must take account of the overall uncertainty of the standard materials themselves and this can vary from around ±1% → ±3%. The error associated with pipetting or weighing of the standard is also pertinent. If external standardization is employed we have to know the error associated with the fitted curve, the error of the external standard ratio of the standards and the unknowns. All in all there are at least six major points at which statistical variation can occur and being able to incorporate and evaluate all of them and bring them into a final error term is a considerable undertaking not only from the programming point of view but also from the workers point of view. To be able to obtain an accurate estimate of DPM from CPM for quenched samples is I believe not easy. Getting an adequate error estimate is almost impossible.

Finally let us consider one area where data processing can save a lot of trouble and embarrassment and this is in detecting for example instrument drift, power failure (and proper recovery), chemiluminescence etc. These phenomena can be detected rather readily and the worker alerted to the problem. The advent of fast microprocessors makes it possible to have such detection equipment built into the spectrometer.

An excellent all-round review of data processing in scintillation spectrometry has been written by Spratt recently (2) and a good example of a well laid out and flexible program has been described by Bowyer and Pearson (3).

REFERENCES

1. C.O. Carroll and T.J. Houser, Int. J. Appl. Radiat.
    Isotop. <u>21</u>, 261 (1970).

2. J.L. Spratt <u>in</u> Liquid Scintillation Counting, Vol. 2,
    p. 245 (M.A. Crook, P. Johnson and B. Scales, Eds.).
    London : Heyden & Son (1972).

3. D.E. Bowyer and J.D. Pearson <u>in</u> Liquid Scintillation
    Counting, Vol. 3, p. 94 (M.A. Crook and P. Johnson,
    Eds.). London : Heyden & Son (1974).

REFERENCES

Discussion

Dr. L.I. Wiebe - University of Alberta (CAN)
Dr. Stanley, you stated earlier that many people only
need CPM because they want only relative values. I can
appreciate that DPM are seldom required but it would seem to
me that most procedures at least in their initial stages
require at least corrected CPM to get meaningful relative
values. Would you care to comment on that?

Dr. P.E. Stanley - Queen Elizabeth Hospital (AUS)
Yes, I would agree with that. I think that probably the
best thing to do is to check the sample channels ratio (SCR)
or the external standard of all your samples if you have a
well behaved cocktail system. Then you might check that all
your samples lie within some channels ratio range that you
consider to be acceptable. In this way you need only to
examine more closely those samples that lie outside that
range.
I think it is a realistic factor that most people do
only want CPM. I think that provided there is some flag
which will alert them to a problem in a sample, if something
is wrong with that sample, this is all that is often required.
This is especially true when routinely counting large numbers
of samples.
I know this is horrifying to a lot of people but this
is a real life situation.

Dr. W. Reid - Saskatoon Cancer Clinic (CAN)
I noticed that Dr. Noakes did not mention electrical
interference as being a problem (with low level counting).
Is that because of the scintillation counter you had, or
were you working in a special lab?

Dr. J.E. Noakes - University of Georgia (US)
This problem did exist in our lab for a long time.
What we finally did to resolve the problem was to design
our own power system where we have a battery pack which
operates under continuous charging from the A.C. current and
can hold five counters up to twenty-four hours in case there
is an outage.

Dr. W. Reid
So you kill two birds with one stone, you get rid of
the power failure problem and the electrical interference.

Dr. J. E. Noakes
We have an isolated power supply system.

Dr. W. Reid
The second question I would like to ask along that line
is have you ever used counters with a sort of variable dead
time? A long time ago I did this, intentionally made the
dead time of the order of one-hundred microseconds to one
millisecond, because often noise of the nature we are
discussing and even some of the background is kind of time
correlated. I have never actually studied this. Have you
studied it?

Dr. J.E. Noakes
You have to incorporate a delay line into your electronics when using a sodium iodide anticoincidence shield. This
is necessary because of the slow scintillation response of
sodium iodide as compared to the liquid scintillators.

Dr. W. Reid
I mean long dead times like one millisecond. In fact,
five-hundred times longer than the response time of sodium
iodide.

Dr. J.E. Noakes
No, we have not extended it out that long. When you do
that of course, you reduce your counting efficiency.

Dr. W. Reid
Not really if we are talking about low level counting.

Dr. J.E. Noakes
For low level counting that is true, but you are
reducing it somewhat. We tried this idea out to extended
times, but not to the degree you suggest, and we did not
see any advantage in extending it out that far.

Dr. H.H. Ross - Oak Ridge National Laboratory (US)
Dr. Noakes, I wanted to ask you if my memory serves
me correctly. Was there not a commercial liquid
scintillation counter that had the phototubes at ninety
degrees. It would seem that this orientation would severely
cut down the crosstalk between photomultiplier tubes.

Dr. J.E. Noakes
     Yes, Tracerlab had one early in the game and the idea
was to place one photomultiplier tube (PMT) so it would not
directly view the other PMT and thereby reduce crosstalk.
But in order to obtain good counting efficiency they put a
mirror at the right angle, and the photons generated from
the corona of one PMT would then be reflected to the second
PMT and the net effect was the same as the one-hundred and
eighty degree geometry.

Dr. D.L. Horrocks - Beckman Instruments (US)
     I would like to make a couple of comments and then
listen to Dr. Noakes' reply.
     We investigated the distribution of background pulses
as a function of the pulse height and we found that we
could quench the liquid scintillator very drastically
without affecting the distribution of the pulse heights
that were coming just from background. Our conclusions
from this was that most of the background was coming from
cosmic events that were interacting with the phototubes
and giving crosstalk between the two phototubes, rather
than cosmic events interacting in the liquid scintillator.
This idea may be reinforced by remembering that the liquid
scintillator is a very low density medium. Some of it
may be due to different cosmic events occurring in the glass
but again those that occur in the glass are not affected by
the quenching that occurs in the liquid scintillation
cocktail itself. That is one point. I wanted to ask if
you had looked at that and what your conclusions were on
that.
     The second point I would like to bring up and ask for
your comment on was have you ever tried to change the high
voltage on your phototubes? One of the things we found was
that we could take a commercial instrument and by decreasing
the high voltage on the phototubes we could, particularly
for [14]carbon counting (for tritium it is almost impossible
to do because efficiency falls off too rapidly with high
voltage) get backgrounds of around four counts per minute
with counting efficiencies of sixty percent. This gives
a figure of merit (E^2/B) of around nine hundred. I am
sure with a little extra shielding and a few things as you
suggested earlier, it could be improved even further.
     Perhaps you would make some comments on your experi-
ences on these two points.

Dr. J.E. Noakes

Yes, there has been some work published on this interaction with the glass, and we did look at this. However,
since we got such a definite response from just diminishing
the volume of our cocktail from say twenty through to five
millilitres while keeping all other variables stable, we
felt that the volume sensitivity was much more a factor.
This was also a factor which we could influence more readily
rather than trying to alter any of the glass components or
trying to get RCA to do something with their phototubes.
Actually we would like to have, instead of RCA's two inch
diameter tube, a one inch diameter tube that would have the
fast response of a RCA 4501-V4.

Now concerning the other point of adjustment of the
high voltage. I should have mentioned earlier that the
phototubes we use are the RCA 4501-V4 tubes with the gallium
phosphide first dynode. These can go up to twenty-four
hundred volts and perhaps even a little more than that.
Most commercial counters today are designed to get as high
a tritium efficiency as is possible, and ^{14}carbon is of
little concern. However, when you are counting ^{14}carbon
you can gain a lot by lowering the voltage, which does
then diminish the crosstalk and hence the background.

Dr. D.L. Horrocks

I would like to bring out another point. When you
reduce the volume of your sample but contain it in the same
size of glass vial, you reduce the ability of light to be
transmitted through it because you are introducing a second
and third glass-air interface. Just reducing the volume
may very well reduce the ability of crosstalk to go from
one tube to the other. We took a vial and filled it up
with water and the background was not reduced in going from
the liquid scintillation cocktail to just water. I do not
think just the volume reduction of the scintillator is the
complete answer.

Mr. B.H. Laney - Searle Analytic (US)

We have examined the pulse height spectrum of background using a dual parameter multichannel analyzer[1]. For
the range of pulse heights corresponding to ^{14}carbon in
fifteen millilitres of scintillator we have identified which
portion of the spectrum is due to crosstalk. We note that
for crosstalk events one of the phototube signals is always
quite small. This is because that light produced within one
of the phototubes behind its faceplate, is attenuated by the
photocathode before it is transmitted to the other phototube.

Thus, there is a high probability that the light detected by one phototube is much lower than that from the other for crosstalk events. This pulse height difference is sufficient to discriminate crosstalk events from scintillation in the sample[2].

As Dr. Noakes pointed out this crosstalk may be reduced by masking off the outer edge of the phototube faces or by tightening up the coincidence resolving time, or by electronic discrimination.

If you employ lesser pulse height analysis in which only the smaller of the two phototube signals is used, these crosstalk events can be eliminated by simply raising the lower level discriminator. However, if the two phototube signals are summed then we cannot discriminate against crosstalk in this way.

---

[1] B.H. Laney in Liquid Scintillation Counting, Vol. IV, p. 74, M.A. Crook and P. Johnson (Eds.), Heyden and Son, London (1976).

[2] B.H. Laney in Organic Scintillators and Liquid Scintillation Counting, p. 991, D.L. Horrocks and C.T. Peng (Eds.) Academic Press (1971).

Dr. B.E. Gordon - Lawrence Radiation Laboratory (US)
When employing the lesser pulse height approach one observes that a quench curve is valid for samples that are both chemically and colour quenched. I would like a rather exact explanation on why it is that Beer's Law overlaps with all the quenching kinetics that govern chemical quenching.

Mr. B.H. Laney
There are several references to describe why the method works[1,2]. I want to make it clear that I have never made a claim that both coloured and chemically quenched samples fall on the same quench curve; it depends on the range of quenching that you consider. Dr. Noujaim, Mr. Ediss and Dr. Wiebe[3] showed that a much higher level of quenching is required to cause divergence of the quench curves when using lesser pulse height analysis when compared to pulse summation.

In simple terms, we are familiar with the idea that loss in counting efficiency is generaly associated with a shift in pulse height spectra to regions of smaller pulse heights. Essentially the problem with pulse summation in this regard is that for samples that are coloured we get a distribution of pulse heights that are spread over a wider range than that expected for that observed counting efficiency. Actually, in Baillie's[4] original article he pointed this out. He suspected that the divergence between the chemical and colour quenched curves is probably due to the fact that we are dealing with a coincidence system, and not a single tube counter. That is what is really happening here.

When we have a heavily coloured sample (and hence a low counting efficiency) then using the pulse summation method we get a much wider range of pulse heights than that obtained for a sample chemically quenched to the same degree. For the coloured sample those events that take place at the edge of the vial close to one phototube will give rise to a large pulse in one phototube and a small one in the other. If these two pulses are now added together the resulting pulse height may well be as large as those pulses obtained from an unquenched sample. By taking only the lesser of the two pulses we get a pulse height distribution that is more representative of the observed counting efficiency.

Since most quench correction methods detect changes in counting efficiency by monitoring shifts in the pulse height distribution of either the sample or some external standard, the argument given above explains why one might expect lesser pulse height analysis to provide some improvement. I hope that answers the question.

---

[1] B.H. Laney in Liquid Scintillation Counting: Recent Developments, p. 455, P.E. Stanley and B.A. Scoggins (Eds.) Academic Press (1973).

[2] B.H. Laney in Liquid Scintillation Counting, Vol. IV, p. 74, M.A. Crook and P. Johnson (Eds.), Heyden and Son, London (1976).

[3] C. Ediss, A.A. Noujaim and L.I. Wiebe in Liquid Scintillation Counting: Recent Developments, p. 91, P.E. Stanley and B.A. Scoggins (Eds.), Academic Press (1973).

[4] L.I. Baillie, International Journal of Applied Radiation and Isotopes, 8, 1 (1960).

Mr. T. Carter - University of Birmingham (UK)
I'd like to ask Dr. Stanley whether he sees any advantages or disadvantages in fitting to a few points in a narrow range of interest, rather than fitting a whole quench correction curve, when using a computerized system.

Dr. P.E. Stanley
Are you trying to say that you should get the curve to pass through the points, or close to them?

Mr. T. Carter
    No, instead of fitting say a fifteen or twenty point curve, fit a curve over say five or six points around the area of interest.

Dr. P.E. Stanley
    No, I can't say that I do.

Mr. B.H. Laney
    I would like to comment on that. Most of the curves plotting counting efficiency versus some type of quench parameter are dependent upon which quench parameter is used, and what the window settings are. There is really no physical reason why these curves should follow a polynomial function.
    As more points are added to a curve, a higher order polynomial is required to fit that curve closely. The higher the order of the polynomial you use, the more careful the user must be in order to be sure that the curve does not pop out between the points. Force fitting at the data points (by using high order polynomials) can cause very large errors between those points.
    I believe Mr. Carter is referring to a spline fit which we found to be very good.

Dr. S. Apelgot - Institut du Radium (F)
    I would like to return to the problem of reporting experimental results in CPM as opposed to DPM. It is true that in biological laboratories people use CPM, but this is only correct if the counting efficiency from the different samples is about the same.
    Also, I have observed errors due to the counting vials themselves. I work mostly with glass vials and I decided to screen them. I found that it was necessary to reject ten to fifteen percent of them because of their deviation from the norm.

Dr. P.E. Stanley
    Yes, I would agree with that comment about glass vials. If you look at the transmission of light through them they are indeed quite variable and 15% would certainly be a number that I would agree with. There is a paper in the latest issue of Analytical Biochemistry which considers light transmission through scintillation vials. (Corredor et al. Anal. Biochem. 70, 624 (1976)). It was particularly about bioluminescence, and they used bioluminescence as a light source. They were getting (if my memory serves me correctly)

273

about ± 10% variation within a batch and quite large differences from batch to batch. This is certainly an important problem. I know of several authors who go through their vials and select them in critical applications.

Dr. W. Reid
 I would like to comment on some of Dr. Stanley's remarks concerning RIA. The sort of problems one gets clinically in radioimmunoassay are almost an order of magnitude more complicated than just quench corrections. However, in my experience the associated calibration curves deviate from a straight line by at most just a few percent. A second order polynomial fits very adequately over the entire range of concentrations we use.
 We do have some problems, however, in estimating the errors in our results when using these calibration curves. By determining the variability in the calibration curves (by generating many of them); and by measuring replicates of our serum samples we can find that the variance is a function of the resulting concentrations. However, we need an off line computer, and an accumulation of these calibration curves for a year or so to be able to do it.

Dr. P.E. Stanley
 The estimation of errors in radioimmunoassay curves is an ever present problem. It seems that there is a special breed of statisticians who make it their business to deal with problems such as this.
 People doing RIA by liquid scintillation counting frequently fail to do quench correction on their samples. Whether they be urine or blood samples, they can contain all sorts of other odds and ends, especially if the patient is on one or more sorts of medication. So you can imagine the sort of problems they get. This may well be why you see a bigger variation in patient samples than that which you see on your standard curve.

Dr. E. Schram - University of Brussels (B)
 My question has to do with the use of delay lines in coincidence circuits used to demonstrate the presence or absence of chemiluminescence. This system was used in at least one instrument more than fifteen years ago and has been reintroduced in recent instruments.
 Is there any reason why it has not been used for many years, and how adequate is it for showing the presence of chemiluminescence?

Mr. B.H. Laney

The delayed coincidence method of determining accidental coincidence has been around for a long time. It is only in recent years when the detection efficiency of a scintillation counter has become high enough for chemiluminescence to again be a problem. When phototubes had low quantum efficiency the problem was of no great concern. Now, especially with surfactant systems, chemiluminescence is an ever-increasing problem. The delayed coincidence method of monitoring or measuring accidentals is an old technique revitalized.

Dr. Horrocks was describing a new system that is included in recent Beckman instruments, and perhaps he would care to comment.

Dr. D.L. Horrocks

In the Beckman system we have tried to approach the problem of measuring the chance coincident rate employing a different technique than the use of a delay line. The reason why we avoided the delay line method is the fact that since radioactive decay events are random in themselves, delaying one signal relative to the other risks a certain probability of having chance coincidence between two different beta events. Thus, as the count rate goes up, the chance coincidence would be increasing only because the sample activity has gone up, not because of any single photon event.

The approach we take involves monitoring the sample prior to counting. After the vial goes down into the counting well, the first thing we do is to turn off the coincidence requirements and measure the count rate in a wide open window. In that wide open window we get a single photon count (SPC_1). That single photon count is going to be the sum of the singles from each phototube (S_1 and S_2), plus the sample activity (S_a) plus the chance coincidence rate (S_c).

$$SPC_1 = S_1 + S_2 + S_a + S_c$$

That rate (SPC_1) is usually a fairly high count rate so we can get good statistics on it in a very short time. When we then go into regular counting of the samples, we want to count in particular counting windows. While this normal counting is going on we have a coincidence pulse circuit which acts as a gate for the pulses coming from the pulse height analyzer before they go to the scaler. This coincidence gate provides a pulse whenever it detects a coincidence event. We count these pulses with a scaler to provide a measure (S_m) of the sample activity plus the chance coincidence.

$$S_m = S_a + S_c$$

If we take the single photon count (SPC$_1$) and subtract S_m from it we are left with $S_1 + S_2$. Now for the single photon rate to contribute to the chance coincidence rate it has got to be something that is occurring in the cocktail itself. The chances are that it is going to be seen equally by both phototubes.

Thus, $$S_1 \simeq S_2 \simeq \frac{(SPC_1 - S_m)}{2}$$

Now we have all the information we need to calculate the chance coincident rate.

$$S_c = 2 \cdot T_r \cdot S_1 \cdot S_2$$

Where T_r is the resolving time of the instrument and thus

$$S_c = 2 \cdot T_r \left(\frac{SPC_1 - S_m}{2} \right)^2$$

Since S_m is a measure of all coincidence events, and since our sample counting windows may accept only a portion of these events, our observed sample counts will be less than S_m. Thus, S_m will be known to equal or greater precision than the sample counts.

If we report $100 \cdot S_c / S_m$ then this factor gives the maximal percentage contribution of chance coincidence to the measured sample countrate.

Another feature of this technique is that the single photon count rate can be measured (SPC$_1$ and SPC$_2$) both before and after the sample counting period. In this way you can tell whether the chance coincidence rate has changed dramatically while counting, or whether it has remained constant.

We do not recommend this as a way of correcting for chemiluminescence because you do not know at what pulse height the chance coincidences occur. You could only use the factor given above to determine what correction should be made to your sample count rate if that sample count rate was accumulated using a wide open window. However, this technique is useful for screening samples that might have a high chance coincidence rate. Then problem samples could be recounted later, or some corrective action taken to remove the chemiluminescence.

Dr. A.A. Noujaim

I would like to ask two questions. Firstly, in view of the problems that have been mentioned with regard to photomultiplier tubes, can I have some estimate from the

manufacturing industry as to how far away we are from
seeing a liquid scintillation counter with a different
light sensitive device?

Secondly, was Dr. Horrocks assuming that chemiluminescence is a single photon event or a multiple photon event
in his calculations?

Dr. D.L. Horrocks

In answer to your first question, we really haven't
seen anything that could replace phototubes at present.

We have been talking to the phototube manufacturers and
we imply that the phototube is the weakest link in the
liquid scintillation process. The quantum efficiency is
only 28%. We keep saying we want more efficiency, but
this development is at a standstill. What really has to
happen is a total breakthrough. Maybe a new idea, a
development of a new photocathode material just like when
the bialkali system was discovered. I do not really see
any big improvements at the present time.

Dr. A.A. Noujaim

What I had in mind was light sensitive diodes.

Dr. D.L. Horrocks

It is my understanding, I may be a little naive here,
that the big push on these diodes has been into the
infrared where they are much more sensitive rather than
into the visible. Dr. Kelly is more of a spectroscopist,
he might have some more comments on that.

Dr. M.J. Kelly - Beckman Instruments (US)

At the present time, the photodiodes are just getting
down into the UV visible regions, and at this stage we do
not see that they will be sensitive enough for the liquid
scintillation process. However, there may be some new
developments fairly soon.

Dr. P.E. Stanley

I just wonder if anybody has tried using the 95% of
the energy that is lost. After all, the scintillation
process itself is only 5% efficient, the other 95%
presumably goes away as heat or something along those
lines. Really the scintillation efficiency is the
weakest link in the chain.

Dr. D.L. Horrocks
   The scintillation process is governed by the methods of
radiation interaction with matter. I think those processes
are pretty well controlled by the laws of physics, and I
just do not see any way to get more energy to be converted
into more excited molecules or ions.

Dr. P.E. Stanley
   I was thinking of the other 95%. Can we measure that
by some other means, not a light detector?

Dr. D.L. Horrocks
   Well, one could always build a very sensitive calori-
meter!
   The second question Dr. Noujaim asked was whether I
considered all the events as single photon events in
calculating the random coincidence rate. The answer is
that it does not make any difference as long as they are
random events because we look in a wide open window. If
it is a chance coincidence between a single event in one
tube and a two or three photon event in another tube, it
does not make any difference because it is just a chance
coincident event. The important criterion is that they
are chance events, and of course most of them will be
single photon events.

Dr. H.H. Ross
   I would just like to say that in spite of the
difficulties there will be new photon detectors used
very shortly. I am confident of that. If you observe
what has happened to the calculator and digital watch
industry you can surmise that photodiodes, charge coupled
devices and charge injection devices are developing
quickly. Right now there are some photodiodes that have
sensitivities approaching those of photomultiplier tubes
but the geometry is very small, and this of course is
a limiting factor. I cannot predict exactly when it is
going to happen, but every day I see new items becoming
available that I would not have expected to be viable.
   In my group, which is involved with new detectors,
we purchased a charge coupled two dimensional array about
a year ago for four-thousand dollars. This was a state
of the art device; it was the hottest thing off the
diffusion oven! Today that device is obsolete. So we
are in a tremendous state of flux in that particular area

and I think that perhaps in ten years you will find some things that you would not have thought possible.

Dr. E.I. Wallick - Alberta Research Council (CAN)
   I would like to know if there is any control on the radioactivity of the materials used in manufacturing counting instruments.

Mr. B.H. Laney
   At Searle there is a program of lot control on the metallic materials that go anywhere near the detectors used in the counting systems. A sample of each lot is evaluated before anything is fabricated from it, and only when it has been proved to have sufficiently low background is approval for its use given.
   The answer is yes.

Mr. E. Polic
   There is lot control on all materials used in the detector section. The materials of greatest concern are aluminum, stainless steel and lead. Yes, there is very good control over these materials, and I think all manufacturers do the same things we do.

Mr. J. Burnham - New Brunswick Power (CAN)
   Dr. Noakes, with all the techniques you described for reducing background in commercial counters, if you go to your five millilitre lead shielded vial and you do all the other things you describe, how low a background can you get with a commercial machine?

Dr. J.E. Noakes
   First of all we must specify which isotope we are considering. With a commercial instrument, if we are considering tritium the following steps will be required. Firstly, your sample must be prepared in such a form that it may be counted with high efficiency. Perhaps it might be converted to benzene or something of this sort. Then secondly, because of the high counting efficiency you can take advantage of going to smaller volumes. In this way you could probably maintain a counting efficiency for tritium of 60% with backgrounds for a five millilitre vial or three or four counts per minute, or maybe even less.
   There may be some other tricks that could be worked out, but at present this is probably the best you could do.

Dr. L.I. Wiebe
  Our work reported at this conference on using methyl
salicylate as a counting medium indicates that the external
standard quench correction method (ESP) is not going to
work very well for this system. From Dr. Horrock's descrip-
tion I suspect that the 'H' number will encounter similar
problems when using this material. Should there not be
some warning about this limitation in the respective
operating manuals?

Dr. D.L. Horrocks
  I am not sure I have a complete answer for you because
I do not think that the 'H' number technique is applicable
to Cerenkov counting or pseudo scintillation counting. The
'H' number technique requires that we be able to produce a
compton distribution that will be predictable, and give a
nice sharp compton edge that we can analyze. When you do
Cerenkov counting you are dealing with a more complex
situation. Only the most energetic Compton scattered
electrons are going to give you any Cerenkov light. The
distribution of those light pulses is going to be very
markedly compromised by the fact that the resolution will
be very poor. I would never recommend that the 'H' number
be used as a parameter for monitoring quench, or determining
efficiencies in other than the usual liquid scintillation
systems.

SAMPLE PREPARATION WORKSHOP

PANEL MEMBERS

Dr. K. Painter — Chairman

Dr. E.D. Bransome — Standardization in Liquid Scintillation Counting

Dr. D.C. Wigfield — Adsorption in Liquid Scintillation

*Dr. W.E. Kisieleski — Sample Oxidation For Liquid Scintillation

Mr. R. Ferris — Sample Solubilization

Dr. C.T. Peng — Chemiluminescence

*Unfortunately, because of an emergency, Dr. W.E. Kisieleski had to leave the conference early. Dr. J.E. Noakes kindly gave us the benefit of his expertise in sample oxidation during the discussion period of this workshop.

SAMPLE PREPARATION IN LIQUID SCINTILLATION
REMARKS OF THE SESSION CHAIRMAN

Kent Painter
Department of Radiology and Radiation Biology
Colorado State University
And
Micromedic Diagnostics, Incorporated
Fort Collins, Colorado 80521

From the early days of liquid scintillation, sample preparation has been the primary nemisis of this elegant analytical tool. One is faced with the reciprocal effects of augmenting sample volume to increase the total radioactivity in the vial (hence, to improve sensitivity and precision) while at the same time the quenching effect of the larger sample size reduces counting efficiency. Thus, workers have adopted sample figure of merit as a means of optimizing sample preparation conditions.

Sample Figure of Merit = Sample Size x Counting Efficiency

One attempts to balance the sample system to achieve the maximum counting rate in the vial. Sample figure of merit is not to be confused with instrument figure of merit (efficiency^2/background) which is used to monitor instrument performance.

From a historical perspective it is interesting to compare aqueous tritium counting data from early workers with data obtained by state of the art methodology. These data are presented in Table 1. Due to limitations of incorporating aqueous samples into primary organic solvents early workers utilized 60-85ml counting bottles to achieve reasonable counting rates and the first commercial counters were designed to accomodate this configuration. By the late 1950's improvements in instrument design, which simultaneously reduced background and improved counting efficiency, and novel cocktail formulations allowed the use of 20ml counting containers which became standard.

The use of tri-functional surfactants of the polyethoxy-alkylphenol series developed in the late 1960's is primarily responsible for the considerable progress which has been made toward increasing aqueous sample volumes while minimizing the effects of quenching. Our suggestion several years ago that

Table 1   A Chronology of Progress in Liquid Scintillation Sample Preparation

Method	Tritium Counting Efficiency	Sample Volume	Figure of Merit	Secondary Solvent
Chen 1958[1]	3.7%	1.5%	6	Ethanol
Davidson 1958[2]	6.0%	20.0%	120	Polyether
Bray 1960[3]	10.0%	22.0%	220	Dioxane
Patterson 1965[4]	10.0%	23.0%	230	Triton X-100
Turner 1969[5]	22.0%	21.0%	462	Triton X-100
Commercial 1972[6]	27.0%	40.0%	1080	Unknown
Commercial 1975[7]	30.0%	50.0%	1500	Unknown

[1]Proc. Exp. Bio. Med., 92, 546 (1958).

[2]See reference 1, p.88

[3]Anal. Biochem., 1, 279 (1960).

[4]Anal. Chem. 37, 854 (1965).

[5]Int. J. Appl. Radiat. Isotopes, 20, 499 (1969).

[6]PCS™,Amersham Searle.

[7]Merit™, Isolab.

improvement in sample preparation techniques was rendering large volume liquid scintillation counters obsolete has recently been proven with the introduction of the first compact, table-top, mini-vial counter by a major instrument manufacturer.

LIQUID SCINTILLATION LITERATURE

Since sample preparation is one of the most troublesome areas of liquid scintillation counting, sundry recipes and remedies have been described for solving particular problems. In that regard it is helpful to have at one's fingertips a bibliography of liquid scintillation literature. Since no previous symposium has presented a comprehensive bibliography, one is provided here with commentary.

SYMPOSIUM PROCEEDINGS

The first liquid scintillation symposium was held in August of 1957 at Northwestern University. The proceedings entitled Liquid Scintillation Counting were published by Pergamon Press (1). It is interesting to review many of the original papers presented at the symposium in an era where instrumentation and sample preparation techniques were very crude. One of our distinguished participants in the workshop session, Dr. Chin-Tzu Peng, presented a paper on Sulphur-35 Liquid Scintillation Counting at that very first symposium.

In 1969 a symposium was held at MIT in Boston and the proceedings were published by Grune and Stratton in a volume entitled The Current Status of Liquid Scintillation Counting (2). We are also very fortunate to have with us in our workshop session, Dr. Edwin D. Bransome, who edited this volume and played a major role in the 1969 symposium.

In 1970 a symposium was held at the University of California in San Francisco and the proceedings entitled Organic Scintillators and Liquid Scintillation Counting were published by Academic Press (3). Dr. Chin-Tzu Peng and Dr. Donald L. Horrocks, who is a plenary lecturer at this symposium, edited the proceedings.

Following the first three symposia in the United States the first of four British symposia organized by the Society for Analytical Chemistry was held in 1970 at the University of Salford. The first United Kingdom symposium was published

by Heyden entitled <u>Liquid Scintillation Counting-Volume I</u> (4).
The following year a symposium was held in Brighton, England
which resulted in <u>Liquid Scintillation Counting-Volume II</u> (5)
and in 1973 a third symposium was held in Brighton which
resulted in <u>Liquid Scintillation Counting-Volume III</u> (6). In
1975 a fourth symposium was held in Bath, England and I believe that the proceedings will shortly be published by
Heyden as <u>Liquid Scintillation Counting-Volume IV</u>.

In 1973 a symposium was organized in Australia by Dr. B. A.
Scoggins and Dr. P.E. Stanley, who is not only presenting a
plenary lecture at this symposium, but is also participating
in the workshop session on instrumentation. The Sydney
Symposium was published by Academic Press in a volume entitled
<u>Liquid Scintillation Counting - Recent Developments</u> (7).

And last but by no means least we have our International
Symposium in Canada in 1976. The proceedings of this sympo-
sium are scheduled to be published later this year by Academic
Press.

TEXTS

In contrast to symposium proceedings, textbooks are presen-
tations by a single author or a single group of authors which
attempt to cover theory and applications of scintillation
counting. The first of these was published in 1963 by
Elsevier and was authored by Dr. E. Schram. This book
entitled <u>Organic Scintillation Detectors</u> (8) was an excellent,
comprehensive text on scintillation counting. We are
fortunate to have Dr. Schram as a member of our scientific
advisory committee for this symposium.

The following year Dr. J.B. Birks authored his monumental
treatise entitled <u>The Theory and Practice of Scintillation
Counting</u> (9). This text is still recognized as one of the
foremost authoritative sources on the theory of scintillation
counting.

Nearly ten years passed before another comprehensive text on
liquid scintillation counting was produced. Evidently four
authors simultaneously saw the need for an updated text in
the area of liquid scintillation counting and books were
published by A. Dyer (<u>An Introduction to Liquid Scintillation
Counting</u> - Heyden) (10), D.L. Horrocks (<u>Applications of
Liquid Scintillation Counting</u> - Academic Press) (11),
Kobayashi and Maudsley (<u>Biological Applications of Liquid</u>

Scintillation Counting - Academic Press) (12), and Neame and Homewood (Liquid Scintillation Counting - Halsted) (13). All of these texts were published in 1974.

JOURNALS

Scientific papers relating to liquid scintillation counting are apt to appear in almost any analytically oriented journal. Many of the older journals, particularly those relating to instrumentation developments, are no longer being published. Currently articles on liquid scintillation counting appear in the International Journal of Applied Radiation and Isotopes, Clinical Chemistry, Health Physics, Analytical Chemistry, Clinica Chimica Acta, and Analytical Biochemistry.

A number of review articles are particularly worthy of note. The first by Davidson and Feiligson appeared in 1957 and it is an excellent review of the literature to that date (14).

A second review covering the period of time between 1957 and 1964 by Rapkin appeared in the International Journal of Applied Radiation and Isotopes in 1964 (15).

A third review by Parmentier and Ten Haaf in the International Journal of Applied Radiation and Isotopes encompasses the literature from 1964 through 1969 (16).

Perhaps one of the most practical and useful series of articles on liquid scintillation counting appeared in a British publication Laboratory Practice in 1973. The series of at least seven installments by L.W. Price describes clearly and concisely theory and applications of liquid scintillation counting. The series is highly recommended to both novices and experts as a very practical guide to the liquid scintillation technique.

COMMERCIAL LITERATURE

A final source of information on sample preparation and counting techniques can be found in the literature published by commercial manufacturers. While one must bear in mind that the purpose of this literature is to guide one to the products of the manufacturer, the recipes and the methods described are generally quite good for the particular application. At one time Beckman, Packard, and Searle Analytic had available rather extensive applications manuals

287

with particular emphasis on sample preparation. One should also not forget that these companies have various experts on liquid scintillation counting on their staffs and most are quite willing to discuss with you on an individual basis particular liquid scintillation problems. On instrumentation one could contact Bart Laney of Searle Analytic and Don Horrocks of Beckman. With regard to sample preparation one could contact Roger Ferris of Amersham Searle, Dr. Y. Kobayashi of New England Nuclear, Mr. Robert A. Chudy of Research Products International and Dr. Murray Vogt of Isolab. These gentlemen have spent countless hours on the telephone with hundreds of scientists to help resolve problems in liquid scintillation.

Regrettably the preceeding paragraph may lead one to believe that all of the problems in sample preparation for liquid scintillation counting have been quite well resolved. Unfortunately this is not true, as a number of rather serious problems still plague users of the liquid scintillation technique. The ones which I see as most prominent are standardization, both in terms of radioactivity standards and a standard definition for the term "liquid scintillation grade", the problem of color quenching, the problem of chemiluminescence and the problem of dealing with samples which are difficult to dissolve in scintillation cocktails. In our workshop session here this afternoon the panel of liquid scintillation sample preparation experts will attempt to address these and other questions in their introductory remarks. Following the introductory remarks we will entertain from the floor any and all questions regarding the sample preparation.

REFERENCES

1. Bell, C.G., and Hayes, F.N., Liquid Scintillation Counting, Pergamon Press, New York, 1958. (1957 Symposium - Northwestern).

2. Bransome, Edwin D., The Current Status of Liquid Scintillation Counting, Grune and Stratton, New York, 1970. (1969 Symposium - Boston).

3. Horrocks, Donald L., and Peng, Chin-Tzu, Organic Scintillators and Liquid Scintillation Counting, Academic Press, New York, 1971. (1970 Symposium - San Francisco).

4. Dyer, A., Liquid Scintillation Counting, Volume I, Heyden, London, 1971. (1970 Symposium - Salford).

5. Crook, M.A., Johnson, P., and Scales, B., Liquid Scintillation Counting, Volume II, Heyden, London,1972. (1971 Symposium - Brighton).

6. Crook, M.A., and Johnson, P., Liquid Scintillation Counting, Volume III, Heyden, London, 1974. (1973 Symposium - Brighton).

7. Stanley, P.E., and Scoggins, B.A., Liquid Scintillation Counting -Recent Developments, Academic Press, New York, 1974. (1973 Symposium - Sydney).

TEXTS

8. Schram, E., Organic Scintillation Detectors, Elsevier, New York, 1963.

9. Birks, J.B., The Theory and Practice of Scintillation Counting, Pergamon Press, New York, 1964.

10. Dyer, A., An Introduction to Liquid Scintillation Counting, Heyden, London, 1974.

11. Horrocks, D.L., Applications of Liquid Scintillation Counting, Academic Press, New York, 1974.

12. Kobayashi, Yutaka and Maudsley, D.V., Biological Applications of Liquid Scintillation Counting, Academic Press, 1974.

13. Neame, K.D., and Homewood, C.A., Liquid Scintillation Counting, Halsted, New York, 1974.

LITERATURE REVIEWS:

14. Davidson, J.D., and Feiligson, P., Int. J. Applied Radiation & Isotopes, 2, 1 (1957).

15. Rapkin, E., Int. J. Applied Radiation & Isotopes, 15, 69 (1964).

16. Parmentier, J.H. and Ten Haaf, F.E.L., Int. J. Applied Radiation & Isotopes, 20, 305 (1969).

STANDARDIZATION IN LIQUID SCINTILLATION COUNTING

E. D. Bransome, Jr., M.D.

Department of Medicine
Medical College of Georgia
Augusta, Georgia 30902

In the almost 25 years of liquid scintillation counting,
the convenience and engineering sophistication of counting
instruments has increased remarkably. This laudable progress
has not however resulted in an equivalent increase in accuracy, because there is still no easy road either to the preparation of samples suitable for liquid scintillation counting,
or to devising appropriate known standards. The use of in-
appropriate standards (or none at all), removes one of the
most important criteria of adequate sample preparation.

By definition, the introduction of an unknown sample into
a scintillation vial containing properly selected scintillators and solvent is equivalent to the introduction of an impurity and therefore a certain degree of quenching: attenuation of the radioactivity itself and of the photoelectrons
resulting from interaction with scintillators (1).

The LS counting literature is already replete with exposi-
tions of how to employ the three approaches to monitoring
quenching by:

1. Internal standardization of each sample
2. Plots of sample channels ratios vs counting efficiency
3. Plots of external standard channels ratios vs counting
    efficiency.

Recent short books by Kobayashi (2) and Horrocks (3) can be
consulted by the novice. Dr. Peng discusses chemiluminescence - the other bane of sample preparation - in this
volume (4).

The loss of radioactivity and decreased fluorescence
quantum yield are my concerns in this brief discussion of how
one decides whether a "quench correction" algorithm is appropriate for a specific sample. For standardization to be
valid, the samples must be both "homogeneous" and similar to
the series of standards used to construct quench-correction
curves.

Homogeneity of the Samples

A homogeneous sample can be defined as one in which the composition of the volume traversed by emitted radiation is the same everywhere (5). True solutions are of course included in this definition, but it encompasses any situation in which the sample is evenly dispersed in a four π system. Absorbtion or adsorption of samples results in a 2π system. None of the three techniques of standardization are capable of monitoring loss of counting efficiency when a sample is in 2π configuration. Dr. Wigfield (6) and our own laboratory (7,8) have been concerned with this problem. Counting samples on solid supports is thus unacceptable.

Gel or emulsion counting may qualify if the droplet or particle sizes are small compared to the length of the pathway of the α, β or γ radiation. A sample "homogeneous" for one isotope may therefore not be so for one of lesser energy. E. B. Muller has proposed a fairly rigorous basic test of "homogeneity" in surfactant systems: that the efficiency of counting ^3H_2O and ^3H-toluene be identical (9). A somewhat less sensitive but more generally applicable test is the "double ratio plot": comparison of a quench correction curve derived using sample channels ratios to one obtained with external standard channels ratios. If the two are discordant for quenched samples the sample is not "homogeneous" (10). It is also obvious but probably important to mention that a significant portion of energetic radioactivity may escape altogether from the counting vial. In such a situation quenching may result in an apparent increase in counting efficiency (11).

Similarity of Standards and Samples

It may seem obvious in the conduct of an international conference on liquid scintillation counting that quenched standards should imitate the composition of the samples to be counted. In practice however, they usually do not. Kalbhen and Rezvanie have shown that different problems were encountered with each of 17 cocktails, 3 solubilizers and 2 sample oxidation methods (12):

1. Chemiluminescence
2. Effects of sample volume on counting efficiencies
3. Phase separation.

Little and Neary have shown that for each sample type and solvent, there is a different optimum concentration of scintillator (13). There are significant differences between color and "chemical" quenching, the magnitude of which depends on the energy of the isotope, the counting equipment, the counting system, the degree of quenching, the electron density of the samples and the colors involved (14).

If surfactants are used they may act as scintillators with ^{14}C or more energetic isotopes; standards not containing similar amounts of the surfactant may therefore be extremely inapropos (15). If the phase separation of an emulsion is unstable or if the sample precipitates out of solution over time (a significant problem with inorganic samples) a 4 π system becomes a 2 π system and no series of standards will be valid. Finally, differences in vials may invalidate quench correction curves obtained with external standardization (16).

Conclusions

a) the quenched series of sealed samples commercially available should seldom be used for anything but calibration of counting instruments.

b) sample "homogeneity" for the isotope being counted should be established, with the caution that changing amounts of the sample may also affect the relationship of the sample counting efficiency to the quench correction curve obtained from a set of standards.

c) low energy radionuclides in heterogeneous systems (eg solid supports) are often not amenable to standardization. The effects of adsorption etc. on pulse height spectra are often unpredictable.

d) standards for quench correction curves should be as similar as possible to the unknown samples to be counted.

e) if sample composition is unknown, or cannot be reproduced for standards, the best tactic is if possible to resort to combustion in sample preparation. Then standards which will approximate unknown samples can easily be devised.

References

1. D.L. Horrocks, this volume.
2. Y. Kobayashi and C.V. Maudsley, in Biological Applications of Liquid Scintillation Counting. New York and London: Academic Press (1974).
3. D.L. Horrocks, in Applications of Liquid Scintillation Counting. New York and London: Academic Press (1974).
4. C.T. Peng, this volume.
5. B.E. Gordon, in Liquid Scintillation Counting, Vol. 3, p. 109 (M.A. Crook and P. Johnson, Eds.). London: Heyden and Son (1974).
6. D.C. Wigfield, this volume.
7. E.D. Bransome and M.F. Grower, Anal. Biochem. 38, 401 (1970).
8. E.D. Bransome and M.F. Grower, in Organic Scintillators and Liquid Scintillation Counting, p. 683. New York and London: Academic Press (1971).
9. E. Bush Muller, in Liquid Scintillation Counting, Vol. 3 p. 47 (1974).
10. E. Bush Muller, Int. J. Appl. Radiation Isotopes 19, 447 (1968).
11. F.G. Winder and G.R. Campbell, Anal. Biochem. 57, 477 (1974).
12. D.A. Kalbhen and A. Rezvanie, in Organic Scintillators and Liquid Scintillation Counting, p. 149 (1971).
13. R.L. Little and M.P. Neary, Ibid., p. 431.
14. M.P. Neary and A.L. Budd, in The Current Status of Liquid Scintillation Counting, p. 273 (E.D. Bransome, Ed.). New York: Grune and Stratton (1970).
15. S.E. Sharpe and E.D. Bransome, Anal. Biochem. 56, 313 (1973).
16. K. Painter and M.J. Gezing, in Liquid Scintillation Counting, Vol. 3, p. 34 (1974).

ADSORPTION IN LIQUID SCINTILLATION COUNTING

Donald C. Wigfield
Department of Chemistry, Carleton University
Ottawa, Ontario, Canada

When a radioactive sample becomes adsorbed to the inner walls of a liquid scintillation counting vial instead of remaining in solution, the count rate obtained is no longer a simple reflection of its activity. Thus, the phenomenon of adsorption, although a subject of interest in many areas of physics and chemistry, is, in the area of practical liquid scintillation counting, an irritating and continuing problem. The following discussion is concerned with the practical aspects of recognizing and overcoming the problem.

Nature of the Problem

As a result of the changed counting geometry of an adsorbed sample, and the back-scattering energy loss, counting losses in the channel of interest occur, coupled with the appearance of pulses in channels of lower energy. This overall effect is therefore identical to that caused by quenching — attenuation of the count rate and the pulse height shift to lower energy — with the following crucial difference: quenching is a property of the solution and so the count rate of an external standard is also controlled by it; adsorption, on the other hand, is a phenomenon concerning only the radioactive solute and not the solution as a whole, and thus the count rate of an external standard is essentially independent of the adsorptive state of the radioactive solute. Although this difference may be exploited (see below), the problem of adsorption is, therefore, a particularly insidious one in the sense that it gives rise to erroneous counting data in such a way that is not obvious in the normal operation of a liquid scintillation counter. As Litt and Carter have pointed out (1), the deceptive simplicity of operating modern counters makes the problem all the more acute.

In view of the pulse height shift caused by adsorption, the problem will clearly be more serious for situations in which narrow channels are employed. $^{14}C-^{3}H$ double-label counting, for example, could suffer from the double effect of loss of counts in the ^{14}C channel, and additional ^{14}C spillover into the ^3H channel.

Detection of Adsorption

In our laboratory, the following three tests have been found useful to establish the question of adsorption or non-adsorption of a particular counting mixture.

1. Dilution with non-radioactive carrier. The idea of this test, extensively discussed by Litt and Carter (1), is to produce radioactive solute of such low specific activity that essentially all the surface sites are saturated with inactive carrier, with the result that the active material is in solution and is counted normally. Thus, for adsorbed compounds the apparent activity increases, while for non-adsorbed compounds the activity remains constant. Two disadvantages of the method are, firstly, that inactive material may not always be available, and secondly, that if the limit of solubility is exceeded, misleading results due to heterogenous counting may be obtained. The possibility that the addition of inactive carrier may cause additional quenching is not, of course, of concern as this is corrected for by normal external standardization.

2. Vial emptying. Adsorption may be detected by emptying the suspected vial, refilling the vial with fresh scintillator solution, and recounting. In our experiments on ^{14}C-labelled compounds in glass vials, there is a dramatic difference between non-adsorbed compounds (99.7% of activity removed from the vial) and adsorbed compounds (over 95% retained in the vial). It is worth noting that a small retention of activity (e.g. 16% for benzoic acid) does not appear sufficient to cause any counting anomalies (2).

3. Measurement of the Adsorption Shift. This method, developed in our laboratories three years ago (3), takes advantage of the fact that adsorption affects a properly chosen sample channels ratio (SCR) but leaves unaffected the external channels ratio (ESR), whereas quenching affects both SCR and ESR. Thus, the normal empirical relationship between SCR and ESR in the absence of adsorption may be derived, and the normal SCR calculated for any value of ESR. The Adsorption Shift is defined as the difference between the calculated and observed SCR. Adsorption gives rise to non-zero values of the Adsorption Shift.

Prediction of Adsorption

Two prime requirements for adsorption problems to occur
are firstly the counting of a compound with adsorptive
properties, and secondly, a sufficiently high specific
activity. An extensive survey of ^{14}C-labelled organic
radiochemicals in glass vials (Amersham-Searle low-level
spectravials) was carried out in our laboratories in 1974
(5). It was found that the types of adsorptive radiochemical were, in fact, very few and essentially limited to
compounds with a high oxygen content. The most prevalent
group of adsorbed molecules were those with more than one
carboxylic acid group, there being an interesting sharp
distinction in this regard between mono- and dicarboxylic
acids. In contrast, several other relatively polar
compounds with high nitrogen content (e.g. barbituric acid,
adenine) were not adsorbed. Litt and Carter (1) have found
a number of amino acids to be adsorbed both on glass and, to
a lesser extent, plastic vials. It is perhaps worth noting
that organic chemists (reaction mechanism studies) are
therefore unlikely to be troubled by adsorption problems,
but that many biochemical studies (amino acids, compounds of
intermediary metabolism) are liable to be plagued by the
problem.

The second requirement for adsorption, that of
sufficiently high specific activity, is clearly due to
carrier saturation of adsorption sites mentioned above. In
studies in our laboratory on glass vials (Amersham-Searle
low-level spectravials) the critical concentration to
saturate the sites was found to be approximately 2×10^{-6} M,
corresponding to 3×10^{-5} mmoles/15 ml of counting solution
and, if the counting activity is 50,000 dpm, to a critical
specific activity of approximately 0.5 to 1.0 mCi/mmole,
below which adsorption is likely to be of rapidly diminishing significance. It should be noted that 3×10^{-5} mmoles
(above) corresponds to <1.0 mg for compounds of molecular
weight under 30,000 so that for compounds in this category,
addition of 1 mg of carrier should eliminate the adsorption
problem, provided that the material is sufficiently soluble.

Overcoming the Adsorption Problem

Two approaches are possible. The first approach is to
change the system in such a way as to eliminate adsorption
of the radiochemical involved. The most obvious way of

achieving this objective is dilution with radioactive carrier to apparently constant activity, a method which is subject to the limitations mentioned above (detection of adsorption). Alternatively, a change in the type of vial or solution may give a non-adsorbed situation; recently, emulsion counting of sodium orthophosphate has been reported to eliminate the adsorption problems of solution counting (6).

The second approach is to make use of the erroneous data obtained from the adsorbed radiochemical. In a study in our laboratory in 1974 on ^{14}C-labelled compounds, it was found that not only is the Adsorption Shift a good detector of adsorption, but also that the magnitude of the Adsorption Shift is reliably related to the extent of counting loss due to adsorption (7). Thus, by setting up an additional calibration curve, it becomes possible to correct for the adsorption counting losses (7). As is the case for external standardization, the calibration curve is applicable only for the instrument settings under which it was constructed.

References

1. G. J. Litt and H. Carter in The Current Status of Liquid Scintillation Counting, pp. 156-163 (E. D. Bransome, Ed.). New York : Grune and Stratton (1970).

2. D. C. Wigfield, Anal. Biochem. 63, 286 (1975).

3. D. C. Wigfield and V. Srinivasan, Int. J. Appl. Radiat. Isotopes 24, 613 (1973). A similar double ratio method for distinguishing 2π from 4π counting has also been suggested by Bush (4).

4. E. T. Bush, Int. J. Appl. Radiat. Isotopes 19, 447 (1968).

5. D. C. Wigfield and V. Srinivasan, Int. J. Appl. Radiat. Isotopes 25, 473 (1974).

6. R. Tykva, Anal. Biochem. 70, 621 (1976).

7. D. C. Wigfield, Anal. Biochem. 59, 11 (1974).

SAMPLE OXIDATION FOR LIQUID SCINTILLATION COUNTING:
A REVIEW

Walter E. Kisieleski and Evelyn M. Buess

Division of Biological and Medical Research
Argonne National Laboratory
Argonne, Illinois 60439, U.S.A.

ABSTRACT

The liquid scintillation spectrometer is a versatile
instrument for the measurement and analysis of low-energy
beta emitters, especially hydrogen-3 (tritium) and carbon-14.
On the other hand, biological materials as well as environmental samples are most difficult to prepare as true solutions for liquid scintillation counting and present unique
problems in sample preparation. To overcome problems of
sample solubility, quenching, and chemiluminescence, a more
universal preparation technique can be achieved if the sample
is burned at red heat in an atmosphere of oxygen and the carbon and hydrogen converted into carbon dioxide and water and
quantitatively dissolved in a scintillator to produce an unquenched sample. A number of methods of oxidizing samples
for liquid scintillation counting are discussed. Experimental studies using carbon-14 and tritium are presented and
potential application to biological and environmental problems are considered.

INTRODUCTION

The number and variety of toxic substances released into
the environment as a result of modern technology increases
year by year. Biological and medical research is aimed at
the solution of the health-related environmental problems in
order to develop understanding of the interrelationships between chemical and physical factors in the environment and
human diseases. Since all organic compounds contain carbon
and most contain hydrogen, the determination of carbon and
hydrogen is the most frequently performed analysis. The liquid scintillation spectrometer is a versatile instrument for
the measurement and analysis of low-energy beta emitters, especially hydrogen-3 (tritium) and carbon-14. However, biological materials as well as environmental samples are most difficult to prepare as true solutions for liquid scintillation
counting and present unique problems in sample preparation.

299

The problems of sample solubility, quenching, and chemiluminescence can be overcome in a simple preparation technique if
the sample is burned at red heat in an atmosphere of oxygen
and then the carbon and hydrogen converted into carbon dioxide and water, respectively, and quantitatively dissolved in
a scintillator to produce an unquenched sample.

In what follows, we shall review the general features of
biological and medical investigations that place demands upon
the design specifications of liquid scintillation counters
and associated methodology. Special emphasis will be given
to the oxidative technique for sample preparation. No attempt is made to review completely all biomedical applications, and most of the examples given are those originating
in our own laboratory.

Biological variability places arduous demands upon a tool
applied to biological problems. Individual members of an apparently homogeneous population of living systems do not all
respond alike to the same stress or treatment, and the limits
of normal response are usually quite wide in comparison with
physical measurements. Such variability usually necessitates
statistical interpretation of results, which calls for large
numbers of determinations. In turn, large numbers of determinations require that the method of measurement be simple
and involve as little sample preparation as possible. Simplicity and speed are often more to be desired than very high
precision and accuracy, since biological variations are frequently the limiting factors in interpretation of results
(1-3).

Also of importance is the wide variation in nature and
composition of biological samples. Biological samples may be
organic or inorganic. They may consist of chemically pure
substances such as amino acids, steroids, and sugars, or complex mixtures such as animal excreta, whole organs and tissues, or suspensions of cells. Their solubility in the usual
nonpolar scintillation solvents may vary from complete misci-
bility to essentially complete insolubility. Samples also
may be completely inert to the scintillation process or they
may be highly effective quenchers (4,5).

Biological investigations also require wide variations
in sample size. Studies of natural levels of radioactivity
or of contamination from worldwide radioactive fallout may
require unusually large samples to supplement the sensitivity
of detection. At the other extreme, the amount of sample

available may necessitate measurements on extremely small samples. Examples of the latter are experienced in equilibration studies between the blood stream and the cerebral spinal fluid, or kinetics studies of the anterior chamber of the eye (6-8).

In summary, biological and medical investigations by nature call for counting systems with the greatest of versatility. Among the requirements are (a) analyses of large numbers of samples with a minimum of processing; (b) high sensitivity; (c) wide adaptability as to variations in sample size; and (d) accommodation of wide variations in nature and chemical composition of the sample. This, of course, is a difficult order for a single counting system to achieve, but liquid scintillation systems have shown considerable promise in being able to meet this challenge.

MATERIALS, METHODS, AND DISCUSSION

Liquid scintillation counting is now undoubtedly the most extensively used method for measuring weak beta-emitting isotopes, and various techniques have been developed for their measurement in the presence of biological tissues (9,10). The simplest of these techniques involves suspending or dissolving the tissue in a suitable scintillation medium. The severe quenching inherent in such methods, caused by the natural color of most biological materials, can be partially overcome by decolorizing with hydrogen peroxide. Despite this modification, however, the weight and nature of tissues that can be assayed by direct dissolution methods are limited.

Tissue, plasma, or urine can be also solubilized in aqueous or alcoholic solutions of alkali, or by quaternary ammonium hydroxide compounds and the resulting digest can be assayed. In samples prepared from such digest, however, color quenching is present to a degree that is dependent upon the heme or pigment concentration in the digested material, temperature and duration of heating during solubilization, and amount of digest material dispensed into each vial. Color may be reduced by an additional step, treatment with 30% hydrogen peroxide at the completion of digestion, or correction for quenching differences between samples may be made by preparation of samples with internal standardization or by other means. Unless the protein content is kept low, alkaline digests manifest phosphorescence which must be permitted to decay or be eliminated by the additional step of acidification before measurement (11,12).

301

A more universally applicable preparation technique for handling diverse biological materials is combustion and oxidation, leading to collection of tritium as water and carbon-14 as carbon dioxide. Several approaches, including wet and dry oxidation, have been developed to achieve this objective. Of these the most widely used procedure involves combustion in an oxygen-filled flask.

Oxygen-flask combustion of organic materials was first conceived as an analytical tool in 1892 by Hempel (13), when he combusted sulfur-containing coals suspended in a Pt basket within an oxygen-filled 10-liter bottle. Sixty years later, Mikl and Pech (14) discovered that oxygen-flask combustion was possible for routine semi-microanalytical analysis of sulfur and chlorine. It was a few years later that Schöniger (15,16) improved and extended the method to general analytical usage for various elements. The technique was first used in 1957 for radioactive measurement of carbon-14 by Gotte et al (17) and since then the method and its many variations have been most widely used (18-22).

The major disadvantage of the various oxygen-flask methods is that the combustion vessel is also the collection vessel for the gaseous products of combustion. Samples of scintillator containing absorbent for measurement must be removed from the collection vessel and thus contain variable amounts of quenching products, especially dissolved oxygen (23,24).

The severe quenching property of oxygen causes considerable attenuation and variation in the counting efficiency of the samples to be measured. Deoxygenation of the samples with nitrogen can be used to achieve equilibration of the sample, but stable counting rates are not achieved in all cases due to oxygen reabsorption. This reabsorption can be eliminated only under stringently controlled conditions.

A prerequisite of the oxygen-flask method for large samples is the need to dry biological materials before burning. If any of the tracer in the samples is present in a volatile form the use of the flask method in its simplest form, in which a number of samples are first dried in a vacuum desiccator and are subsequently burned in a combustion flask, is precluded. Such samples with volatile tracers must be processed individually. Existing oxygen train combustion procedures have been adapted by Dobbs (25) in a novel oxygen steam method that is applicable to freshly dissected tissues containing volatile tritiated compounds. In this method, oxygen

at low pressure is passed through a radiofrequency field and
is streamed over the undried tissue. Volatile components of
the tissue are rapidly removed. At the same time, the oxygen
ion reacts with organic material of the tissue leaving an
inorganic residue. The oxidation products containing tritium
are removed from the gas stream in a trap cooled with liquid
nitrogen, and are subsequently dissolved in a liquid scintillator injected into the trap. The method could be a useful
addition to the procedures in current use for assays of tritium in animal tissue because of the absence of memory effects,
the absence of activity in the tissue residue, high trapping
efficiency, and quantitative recovery of tritium from a maximum of 0.8 g of wet tissue. The method has not been adapted
for isotopes other than tritium but does not lend itself to
analysis of dual-labeled compounds. There is also a good
deal of time and tedious labor associated with the method.

Standard vacuum-line combustion techniques have also been
applied to similar problems with limited success; although
these methods offer advantages in accommodating increased sample size, they are time-consuming and require elaborate equipment. They are also subject to serious cross contamination
errors or "memory effects" due to highly absorbent oxidizing
materials present in the system (26).

In our own laboratory a combustion apparatus was designed
and constructed to combine the simplicity of the oxygen-flask
method and the reproducibility and efficiency of vacuum-line
techniques, including the convenience of a direct in-vial collection of the final products of combustion for direct liquid
scintillation counting (27). This method has been referred to
as the vacuum-line method.

Recovery values for carbon-14, sulfur-35, and tritium
from different biological materials, with sample weights up
to 250 mg of dried tissue material, average 98 percent and are
reproducible and independent of the amount and nature of the
material analyzed. The separation efficiency with double-
labeled samples gives recovery values similar to those obtained when only a single isotope is measured.

In measuring double-labeled samples, only a single counting channel is necessary, since the isotopes are chemically
separated. This means instrument settings and calibration
procedures are not critical and can be optimized for each isotope. With these conditions, the range of ratios of one isotope to another that can be determined is greater than that
possible by methods based on differences in energy spectra.

Continuing developments in combustion techniques have retained the oxygen combustion method, but have adopted classical microanalytical tube combustion methods, in which samples burn in a stream of oxygen. This more ideally meets the need for a method suitable for large-scale use (28-30).

The apparatus developed and designed by Peterson et al (31) evolved from earlier tube combustion methods. It collects tritiated water of combustion in cold solvent directly into a scintillation vial. At least 500 mg of dry sample weight can be accepted. The samples are contained in combustible capsules, allowing for a variety of sample types. By this method one investigator can prepare a sample ready for tritium counting every 3 minutes. According to the authors, the method has a collection recovery of 96%, is calibrated by internal standards, and shows a coefficient of variation of 2.0%. Peterson also described a carbon dioxide collection accessory for the previously reported tritium combustion system (32,33).

The utilization of vacuum-line techniques in combination with the basic Schöniger oxygen methodology has evolved into what many term the "second generation of oxygen combustion techniques." Dr. Niilo Kaartinen, at the University of Turku, Finland, developed an automated sample combustion technique incorporating these principles (34). His design has been commercially developed by the Packard Instrument Company of Downers Grove, Illinois. Recently, other manufacturers have produced similar apparatus (35-38).

With available instruments, it is possible to combust either tritium or carbon-14 samples, individually, or double-labeled samples containing both nuclides. The nuclides, in all cases, are collected in separate vials ready for counting. Separation of tritium and carbon-14 in this way allows optimum counting channels to be used in the spectrometer, therefore optimizing counting efficiency and eliminating tedious calculations correcting for carbon-14 spillover into the tritium channel (39-41).

These instruments provide recoveries of 97-98% for most biological samples with dry sample weights up to 500 mg; the precision of recovery has a standard deviation of ± 1%. From initial evaluation reports it appears that these instruments fulfill the many requirements to idealize combustion as a technique for sample preparation of biological materials for liquid scintillation counting (42-44).

304

Biological, medical, and environmental studies are, to a large extent, concerned with how labeled compounds or molecules are metabolized, or incorporated into other substances or into cellular structures. The use of oxidative techniques to prepare samples for liquid scintillation counting for the detection and measurement of the low-energy beta emitters such as tritium and carbon-14 has extended the sensitivity of following biological processes by more than a thousand times over conventional analytical chemistry techniques.

ACKNOWLEDGMENTS

Work supported by the U. S. Energy Research and Development Administration.

REFERENCES

1. C.H. Wang and D.L. Willis. Radiotracer Methodology in Biological Science. Englecliff, New Jersey : Prentice-Hall (1965).

2. W.R. Hendee. Radioactive Isotopes in Biological Research. New York : J. Wiley and Sons (1973).

3. W.E. Kisieleski in Liquid Scintillation, chap. 12. Fullerton, California : Beckman Instrument Company (1971).

4. Y. Kobayashi and D.V. Maudsley. Biological Applications of Liquid Scintillation Counting. New York : Academic Press (1974).

5. E.D. Bransome, Jr., Seminars in Nuclear Medicine $\underline{3}$, 389 (1973).

6. A.A. Moghissi, E.W. Bretthauer, E.L. Whittaker and D.N. McNelis, Int. J. Appl. Radiat. and Isot. $\underline{26}$, 339 (1975).

7. M.E. Hinkle, U. S. Geol. Surv. Prof. Pap. No. 750-B, B171 (1971).

8. V. Nuti and B. Bacci, Farmaco. Edizione Pratica. Pavia. $\underline{27}$, 381 (1972).

9. J.B. Birks. The Theory and Practice of Scintillation Counting. New York : Macmillan and Co. (1964).

10. E.F. Polic in Instrumentation in Nuclear Medicine, Vol. I, p. 181 (G.J. Hine, Ed.). New York : Academic Press (1967).

11. E. Rapkin in Instrumentation in Nuclear Medicine, Vol. I, p. 181 (G.J. Hine, Ed.). New York : Academic Press (1967).

12. P.E. Stanley, Atomic Energy in Australia 13, 29 (1970).

13. W. Hempel, Z. Angew. Chem. 1892, 393 (1892).

14. O. Mikl and J. Pech, Chem. Listy. 46, 382 (1952).

15. W. Schöniger, Mikrochim. Acta. 1955, 123 (1955).

16. W. Schöniger, Facts and Methods for Scientific Research 1, 1 (1960).

17. H. Gotte, R. Kretz and H. Baddenhauser, Angew. Chem. 69, 561 (1957).

18. W. Mertz, Mikrochim. Acta. 1959, 640 (1959).

19. A.R. Britt and W.E. Kisieleski in Abstract of Papers - 167th ACS Meeting. Abst. #41. Los Angeles (1974).

20. H.W. Knoche and R.M. Bell, Anal. Biochem. 12, 49 (1965).

21. G.N. Gupta, Microchem. J. 13, 4 (1968).

22. R.G. Kelly, E.A. Peets, S. Gordon and D.A. Buyske, Anal. Biochem. 2, 267 (1961).

23. W.D. Conway and A.J. Grace, Anal. Biochem. 9, 487 (1964).

24. L.M. Hunt and B.N. Gilbert, Int. J. Appl. Radia. and Isot. 23, 246 (1972).

25. H.E. Dobbs and G.M. Land in Int. Conf. Radioactive Isotop. Pharmacol., p. 121 (P.G. Waser, Ed.). London : Wiley-Interscience (1969).

26. D.A. Kalbhen in Sym. on Liquid Scintillation Counting, Vol. I, p. 149 (A. Dyer, Ed.). London : Heyden and Sons, Ltd. (1970).

27. L.G. Huebner and W.E. Kisieleski, Atompraxis 16, 1 (1970).

28. B.S. McEwen, Anal. Biochem. 25, 172 (1968).

29. S. Mlinko, E. Fischer and J.F. Diehl, Zeit. Anal. Chem. 261, 203 (1972).

30. S. Von Schuching and C.W. Karickhoff, Anal. Biochem. 5, 93 (1964).

31. J.I. Peterson, F. Wagner, S. Siegel and W. Nixon, Anal. Biochem. 31, 189 (1969).

32. J.I. Peterson, Anal. Biochem. 31, 204 (1969).

33. T.R. Tyler, A.R. Reich and C. Rosenblum in Organic Scintillators and Liquid Scintillation Counting, p. 869 (D.L. Horrocks and C.T. Peng, Eds.). New York : Academic Press (1971).

34. N. Kaartinen in Packard Technical Bulletin #18. Downers Grove, Illinois (1969).

35. E. Rapkin and A. Reich, Amer. Lab. Oct, 35 (1972).

36. D.W. Sher, N. Kaartinen, L.J. Everett and V. Justes, Jr. in Organic Scintillators and Liquid Scintillation Counting, p. 849 (D.L. Horrocks and C.T. Peng, Eds.). New York : Academic Press (1971).

37. J.E. Noakes and W.E. Kisieleski in Liquid Scintillation Counting Recent Developments, p. 125 (P.E. Stanley and B.A. Scoggins, Eds.). New York : Academic Press (1974).

38. R.J. Harvey Instrument Corporation, Advertising Literature. Hillsdale, New Jersey (1975).

39. H.W. Hilton, N.S. Nomura and S.S. Kameda, Anal. Biochem. 49, 285 (1972).

40. T. Fujimori, T. Takesue and K. Ishikawa, Chem. Abstr. 79, 21370 (1973).

41. S. Baba, Y. Baba and T. Konishi, Anal. Biochem. 66, 243 (1975).

42. L.H. Scroggins, J. Assoc. Off. Anal. Chem. 56(4), 892 (1973).

43. R.A. Zaroda, Clin. Chem. $\underline{15}$, 555 (1969).

44. J.B. Ragland $\underline{in}$ The Nucleus #20. Des Plaines, Illinois :
     Nuclear Chicago Corp. (1966).

TISSUE SOLUBILIZATION

R. Ferris
Amersham/Searle Corp.
2636 S. Clearbrook Drive
Arlington Heights, Illinois 60005

The problem being considered here is that of finding or devising chemical reagents and methods of application which will allow the solubilization or digestion of plant and animal tissues so that they can be successfully incorporated into liquid scintillation counting mixtures. The tissue may be whole, homogenized, macerated, or in some other state of subdivision. The methods employed should result, ideally, in digested samples which when added to the scintillation mixtures yield clear, colorless, uniform liquids, exhibiting a minimum of quench, a minimum of chemiluminescence, and a maximum of counting stability. All reagents should be inexpensive, easy to handle, and relatively nonhazardous. Reagents should be capable of digesting large size samples, should be capable of rapid and complete digestion, and should not require great expertise in use. Reagents should be as versatile as possible with respect to the types of tissues for which they can be used, and methods of digestion should allow accurate determination of radionuclide content with minimal systematic error. Finally, although it is not possible to achieve all these desirable characteristics with any given reagent or method, it is possible in many cases to design a reagent method which will be optimum for a given sample and set of experimental conditions. Some examples of solubilizing agents and procedures for their use will be given in the following text.

Methanolic Potassium Hydroxide (1)
This reagent was among the first used for solubilization of whole animal tissues. It has the advantages of low cost and rapid digestion for a variety of animal tissues; however, it suffers the disadvantages of limitation of sample size, low counting efficiencies (particularly for tritium), has a potassium-40 background which may prove a problem for low activity samples, and may not always completely digest a sample.

Colorless, Concentrated Nitric Acid (2)

In general, this reagent has the same low cost and rapid digestion advantages and the same low sample capacity and low counting efficiency disadvantages as potassium hydroxide. Examples of tissues it has been used to digest are shaved rat skin (2), and rat brain, skeletal muscle, stomach, liver, and spleen (12).

Formamide (3)

A variety of tissue types have been digested with this reagent including animal epidermis. It has the property of being a strong tissue solubilizer; however, it also is a strong scintillation quencher. It may prove advantageous to use where a sample combines difficulty of digestion and a high-energy nuclide to be assayed.

Sodium Hydroxide Solution (4)

In the reference cited above, the investigator reported that he found it necessary to add Cab-O-Sil to his sample to prevent a constant decrease in counts over a period of ten hours. Stability may be a problem when using this reagent. In general, sodium hydroxide is a less desirable digestion agent than potassium hydroxide because of the reduced solubility of sodium salts compared with potassium salts.

Quaternary Hydroxides and Chlorides

Many references exist on the use of these types of solubilizers. The general method for digesting a sample is to mince, grind, or macerate the wet, whole tissue in a vial. The solubilizer is added and the mixture is digested at a moderately elevated temperature until the sample appears homogeneous. This procedure may take up to several days for solid tissue samples. The next step (optional) is to partially neutralize the digested sample before the scintillator is added. This usually gives higher counting efficiencies, lower backgrounds and improved sample clarity. The disadvantages of these types of solubilizers are slow digestion, high cost, and color-producing reactions with some scintillators. Advantages are relatively high counting efficiencies, and large sample capacity. In general, most basic solubilizers tend to produce chemiluminescence. Quaternaries are particularly prone to this problem. Many times acidification of the digest will reduce or eliminate this phenomenon. When compared with other solubilizers, the quaternaries offer the most versatility with respect to the variety of tissue types which they can be used to solubilize.

310

However, their high cost and other disadvantages must also
be considered if an optimal choice of solubilizer is to be
made for given sample type. Although quaternaries are used
primarily for animal tissue solubilization, they can be used
with soft plant tissue (6). They are unsuitable, however,
for digesting the hard parts of plants which remain undissolved even after long digestion periods. Examples of
commercial quaternary solubilizers are Hyamine-10-X (Rohm
and Haas), Protosol (New England Nuclear), NCS (Amersham/
Searle), and Soluene (Packard).

Perchloric Acid and Hydrogen Peroxide (5)
    Protein was digested and the heme completely decolor-
ized by this combination of reagents. A clear solution
results which can be incorporated into an appropriate
counting solution. This method is claimed to be useful for
digesting tissue labelled with a variety of nuclides. One
investigator (6) found this reagent gave good results when
it was used for digesting dried and hard plant material.
Another investigator (9) using this reagent digested rat
tissue labelled with transuranium elements.

Concentrated Nitric Acid, Perchloric Acid, and Magnesium
Nitrate (7)
    This reagent can be used for oxidizing sulfur-35
labelled tissue to magnesium sulfate. The oxidation products are then dissolved in glycerol and the mixture diluted
with ethyl alcohol and N,N-dimethylformamide. A modification (6) of this reagent gave good results for routine
analyses of dried and hard plant material.

Nitric Acid and Perchloric Acid (8)
    This mixture was used to digest a variety of plant and
animal tissues. Also it is claimed that it can be used for
sample preparation of tissues labelled with a variety of
radionuclides.

Sodium Hydroxide, Distilled Water, Methanol and Triton X405
(10)
    Mammalian tissue was digested with this reagent,
neutralized with nitric acid, and counted in a toluene/
triton scintillator.

Sodium Hydroxide and Bio-Solv BBS-2 (11)
    This combination of reagents is reported to give rapid
solubilization of solid animal tissue. The tissue is pre-
treated with 1 N sodium hydroxide and then neutralized with

the BBS-2 solution. Although the sample size is limited for
this method, relatively high efficiencies are attainable and
the reagents are inexpensive. Bio-Solv BBS-2 is a Beckman
Corp. product.

References

1.   C.P. Petroff, H.H. Patt, P.P. Nair, Int. J. Appl. Rad.
     Isotope 16, 599 (1965).
2.   R.D. O'Brien, Anal. Biochem. I, 251 (1964).
3.   R.C. Meade, R.A. Stiglitz, Int. J. Appl. Rad. Isotopes
     13, 11 (1962).
4.   R. Tye, J.D. Engel, Anal. Chem. 37, 225 (1965).
5.   D. Mahin, R. Lofberg, Anal. Biochem. 16, 500 (1966).
6.   S.K. Das, Indian Soc. for Nuclear Techniques in Agri-
     culture and Biology, Newsletter (Sept. 1973).
7.   W.M. Shaw, J. Agric. and Food Chem. 7, 12 (1959).
8.   H. Jeffay, Anal. Chem. 32, 307 (1960).
9.   A. Seidel, J. Volp, Int. J. Appl. Rad. Isotopes 23, 1
     (1972).
10. J. Dent, P. Johnson in Liquid Scintillation Counting,
     Vol. 2, Chap. 11. (M.A. Crook and P. Johnson, Ed.).
     New York and London: Heyden (1974).
11. A. Stevens, Anal. Biochem. 37, 1 (1970).
12. M. Pfeffer, S. Weinstein, J. Gaylord, L. Indendali,
     Anal. Biochem. 39, 46 (1971).

CHEMILUMINESCENCE

C. T. PENG
Department of Pharmaceutical Chemistry
School of Pharmacy
University of California
San Francisco, CA 94143, U.S.A.

Abstract.   Factors causing chemiluminescence in liquid
scintillation system are briefly reviewed and means of
avoiding and suppressing it given.   Evidence is presented
to implicate singlet oxygen as a causative agent in chemiluminescence.

1. Introduction

Chemiluminescence and photoluminescence are sources of spurious events in liquid scintillation counting. These phenomena were first studied by Lloyd and his coworkers [1] and by Herberg [2]. Since then many reports have dealt with the avoidance and correction of these spurious counts in the measurement of sample radioactivity.

Although chemiluminescence and photoluminescence differ in their mechanism, they often appear to be closely linked in a liquid scintillation system. For example, the counting vials may be intensely photoluminescent while the sample may be either chemiluminescent or photoluminescent or both. Therefore, it may be useful to review briefly the photophysics of energy absorption.

2. Photophysics of Energy Absorption

Upon absorption of energy a molecule becomes excited and may emit light as luminescence or degrade the absorbed energy as heat upon de-excitation. Depending upon the origin of excitation, the luminescence may be classified as photoluminescence, sonoluminescence, chemiluminescence or thermoluminescence (Table I). Among these, the phenomena of photoluminescence of molecular crystals and scintillation solution systems have been employed to advantage for the detection of ionizing radiation. Salient examples are the use of sodium iodide crystals activated with thallium for measuring γ-ray radiation and the application of liquid scintillation systems for detecting all nuclear emissions. The time resolution requirement in pulse analysis has allowed only molecular systems which emit fluorescence with short decay times to be useful for nuclear detection. In liquid scintillation counting, fluors with lifetimes ranging from 96 nanoseconds for naphthalene to 1.1 nanoseconds for PBD [3] are available.

The photophysics of molecules upon absorption of light quanta or radiation has been treated by Birks [4] and others [5,6]. The absorption of light excites the molecule in which an electron is promoted from the highest bonding orbital to the lowest antibonding orbital. For aromatic hydrocarbons, such a transition is a π- π^* transition.

Once excited, the molecule in the excited state must dissipate its energy by internal conversion as heat or by radiative transition with light emission as prompt fluorescence or by intersystem crossing to yield the triplet state which then decays by emission of phosphorescence, delayed

Table I. Types of Luminescence

1. Chemiluminescence (chemical energy)

    Bioluminescence
    Electrochemiluminescence

2. Photoluminescence (radiation energy)

    Fluorescence – Prompt fluorescence

              Delayed fluorescence

                 E-type (thermal repopulation)
                 P-type (triplet-triplet annihilation)

    Phosphorescence

3. Sonoluminescence (thermal energy)

    Triboluminescence

4. Thermoluminescence (radiation and thermal energy)

fluorescence or heat or by dissociation into smaller molecular fragments or by chemical reaction with other molecules. The radiative transition from an excited singlet to the ground singlet state is spin-allowed and the light emission is termed fluorescence and is short-lived (10^{-9}-10^{-6} s), whereas the transition between the triplet state and the ground singlet state is spin-forbidden; the light emission is long-lived (10^{-2}-10^{2} s) and is known as phosphorescence. Owing the long lifetime, phosphorescence is more susceptible to quenching than fluorescence and is only observed in solids, frozen solutions or highly viscous liquids. The spectral maxima of absorption, fluorescence and phosphorescence spectra occur at increasing wavelength in an inverse relationship to their respective energies.

The excited molecules may also undergo energy transfer to other molecules by mechanisms of radiative, collisional, resonance, and exchange transfer. Resonance transfer operates by long-range dipole-dipole interaction (Förster mechanism) and can occur at molecular distances over 50-100 Å between donor and acceptor molecules. It is the predominant mechanism of energy transfer between solvent and solute molecules in liquid scintillation systems [7]. At high concentrations of solute, energy transfer is by short-range exchange forces [8]. Radiative transfer also occurs between the solvent and the solute molecules in liquid scintillation systems [9]. If the acceptor molecule has a higher quantum efficiency than the donor molecule the light output of the system will be enhanced. On the other hand, if the acceptor molecule converts the excitation energy into molecular vibrations which are eventually degraded as heat, the light yield will be decreased, and in such cases, the acceptors act as quenchers. In liquid scintillation systems, efficiency depends upon energy transfer; if the excitation energy transfer is terminated at the quencher molecule, efficiency will be reduced. If sources of excitation other than nuclear emissions are present and result in chemiluminescence, photoluminescence, electrostatic charges, etc., spurious counts will be observed.

3. Mechanisms of Organic Chemiluminescence in Solution

Many chemical reactions emit light but the quantum yields are low and the systems inefficient. With the use of more sensitive light detecting devices, an increasing number of organic reactions are found to be chemiluminescent [10,11].

Chemiluminescence is presumed to comprise one-photon events and can be discriminated against by coincidence gating when measuring sample radioactivity. Chemiluminescence is not subject to photo-reactivation.

The mechanism of organic chemiluminescence in solution involves three stages (i) preliminary reactions to provide the key intermediate, (ii) an excitation step in which the chemical energy of the key intermediate is converted into electronic excitation energy, and (iii) fluorescence emission from the excited product formed in the chemical reaction [10-13]. In reactions in which a fluorescent compound is added to enhance the chemiluminescent emission, an efficient energy transfer occurs and the resulting chemiluminescence is known as "sensitized" chemiluminescence. Chemiluminescence observed in liquid scintillation systems probably belongs to this category.

The mechanism by which chemical energy is provided for chemiluminescent reactions may be by (i) peroxide decomposition [14], (ii) electron transfer by cation-anion radical combination [15] and (iii) energy transfer from excited oxygen molecules and molecular pairs [16]. Some illustrations are given below:

Chemiluminescence derived from peroxide decomposition may be exemplified by the oxidation of oxalyl chloride or oxalic ester with hydrogen peroxide in the presence of 9,10-diphenyl-anthracene [17] which fluoresces at 400 nm and requires at least 71.5 kcal of energy for excitation. The key intermediates formed in these oxidation reactions are presumably monoperoxyoxalic acid HO.CO.CO.OOH and 1,2-dioxetanedione $O=C-C=O$. The latter has been identified by mass spectrometric measurement [18]. Decomposition of the key intermediate through a concerted process in which the cleavage of 2 or 3 bonds occurs simultaneously, provides instantaneous energy release for the electronic excitation of the fluorescent compound [19]. The peroxy acids and 1,2-dioxetanediones are key intermediates involved in the oxidation of triphenyl-imidazole (lophine) bis-acridinium compounds (lucigenin) and many others.

The annihilation of positive and negative aromatic hydrocarbon radical ions generated by electrode processes gives rise to electro-chemiluminescence [15]. Oxidation and reduction of radical anions and cations with chemical oxidants and reductants, respectively, also lead to chemiluminescence. In this reaction, the excited singlet state of aromatic hydrocarbon is formed by electron transfer from the anion radical to the cation radical. Representative examples are phenanthrene and rubrene. Recent investigations indicate a pre-annihilation electro-chemiluminescence involving the formation of the triplet state by energy transfer [20,21].

Reaction in an alkaline solution of hydrogen peroxide with chlorine gas or hypochlorite ion generates singlet oxygen

317

($^1\Delta_g$) which gives rise to red chemiluminescence. When anthracene, acridine, eosin, fluorescein, quinine sulfate or aesculin are present in the solution reaction, H_2O_2 + ClO^-, a strong sensitized chemiluminescence of these compounds is observed. As the energy gap between the ground state and the first excited state in these compounds is greater than the electronic excitation energy from singlet oxygen ($^1\Delta_g$, $^1\Sigma_g^+$), it is infeasible to effect an energy transfer from the latter to the former. Khan and Kasha [16] have shown the occurrence of the excited oxygen molecular pairs ($2[^1\Delta_g]$, $[^1\Delta_g + {}^1\Sigma_g^+]$, $2[^1\Sigma_g^+]$) and their transition to the ground state can provide sufficient electronic excitation energy for the fluorescent compound in the reaction solution. The electronic energy levels of singlet oxygen and excited oxygen dimers are shown in Fig. 1.

Singlet oxygen is present in the air and can be generated from the reaction of sodium hypochlorite with hydrogen peroxide [22], from thermal decomposition of many organic ozonides [23], epidioxides [24], and linear hydrazides [25], from the reaction of potassium superoxide in water [26], and from microwave discharge through gaseous oxygen [27]. Singlet oxygen is also produced by biological and enzyme systems, such as xanthine oxidase [28], rat liver microsomes, NADPH and O_2 [29], human polymorphonuclear leukocytes upon phagocytosis [30] and adrenodoxin-reductase-adrenodoxin enzyme system [31].

4. Chemiluminescence in Liquid Scintillation System

Chemiluminescence is observed when an alkaline tissue solubilizer is added to a dioxane-based scintillator containing naphthalene [32]. It occurs to a lesser extent when the solvent dioxane is replaced with toluene. Treatment of the scintillation solvents, p-dioxane and toluene, by shaking with H_2O_2 increases the intensity of chemiluminescence. Alkaline tissue digest bleached with benzoyl peroxide or H_2O_2 causes more chemiluminescence than unbleached digests. These observvations focus attention on the culprit role of peroxide and excited oxygen.

The necessary conditions for chemiluminescence in some liquid scintillation systems appear to be the presence of peroxide and an alkaline tissue solubilizer such as Hyamine, Soluene, NCS, Protosol, etc. These cannot be used to explain the occurrence of chemiluminescence in emulsion-type liquid scintillators upon addition of water [33] and in samples containing tissue digests treated with perchloric or nitric acid [34].

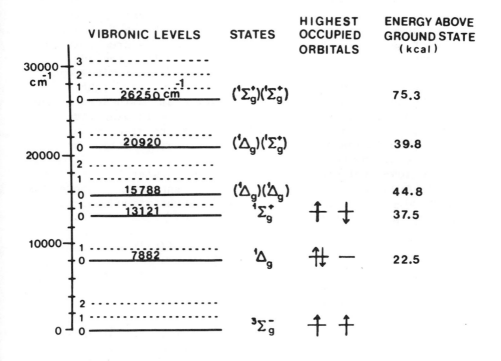

Figure 1. Electronic energy levels of singlet oxygen and excited singlet oxygen dimers (from Reference [26]).

The chemiluminescence-emitting species in liquid scintillators has not been identified. Because of the nature of
the system, fluorescence emission may be due to either direct
or sensitized chemiluminescence or both; fluorescent components in the system may participate in chemical reactions as
well as in energy transfer reactions. In this connection, it
may be mentioned that benzoyl peroxide causes more intense
chemiluminescence than H_2O_2 because the reaction product of
the former can participate in energy transfer more efficiently
than H_2O.

5. Decay Rate, Pulse Height, Temperature Effect and Other Factors Affecting Chemiluminescene

Kalbhen [32,34,35] and Kearns [36] have measured the
chemiluminescence spectrum in liquid scintillation systems.
The pulse height spectrum changes with time and is related to
the decay of chemiluminescence which is dependent upon the
rate of chemical reaction and the lifetime of the fluorescing
molecular species. Initially, the light pulses are of large
amplitude and grow in intensity as the reaction proceeds.
With the passage of time, the intensity of chemiluminescence
diminishes with a concomitant shift of the pulse height towards the low energy end. At low-level chemiluminescence, the
spectrum has a pulse height distribution overlapping the ^3H
spectrum, thus rendering futile any effort to discriminate
against chemiluminescent by adjusting the bias setting or
by the use of channels-ratio method.
The low-level chemiluminescence may persist for many
hours or even days to yield a count rate appreciably above the
background. The decay is temperature dependent and proceeds
faster at elevated temperature.

6. Means of Avoiding and Suppressing Chemiluminescence

Chemiluminescence may be diminished or eliminated by
acidification of the quaternary-base-solubilized tissue digests
with acetic or hydrochloric acid to a pH below 7.0 before addition of liquid scintillator. Oxidizing acids ($HClO_4$, HNO_3) or
sulfuric acid (H_2SO_4) should not be used for neutralization
because they may cause precipitate formation or high colora-
tion and may not always succeed in eliminating the chemiluminescence; in fact tissue samples digested with $HClO_4$ or
HNO_3 have been shown to luminesce [34]. Neame [37] recommends
that the acidification and dilution of alkaline tissue digests
be carried out in sequential steps followed by heating at

50°C and the final addition of liquid scintillator, in order
to avoid chemiluminescence.

The presence of oxygen and peroxide in the scintillator
or solvent may be scavenged by adding minute amounts of ascorbic acid or di-t-butyl-4-hydroxy-toluene (BHT) in combination with hydrochloric acid [36]. The enzyme catalase has
been used to remove residual benzoyl or hydrogen peroxide in
bleached tissue digests [38]. Hydrochloric acid and ascorbic
acid have also been used to decompose residual H_2O_2 in
bleached hemolyzed and jaundiced samples used in digoxin and
T-3 uptake assays.

Affected samples may be stored at elevated temperature
prior to counting to allow chemiluminescence to decay, thereby minimizing interference with normal counting. Use of a
refrigerated spectrometer for sample counting may also serve
to suppress chemiluminescence due to the slow rate of chemical reactions at low temperatures [36]. Light pulses from
chemiluminescence may be discriminated against electronically
by delayed coincidence; the photon monitor in the newer
liquid scintillation spectrometer is based on this principle.
Sample preparation by combustion may be practiced if all
other means of eliminating chemiluminescence fail.

7. A Mechanistic Interpretation of the Cause of Chemiluminescence in Liquid Scintillation System.

Several unique features are associated with the phenomena of chemiluminescence in liquid scintillation system:
(i) the liquid scintillator provides an ideal system for sensitized chemiluminescence to occur because of its highly
efficient energy transferring process, (ii) the persistence
of low-level chemiluminescence over a long duration indicates
either a very slow or a continuing chemical reaction, (iii)
the presence of peroxides is essential, and (iv) the chemiluminescence is enhanced in alkaline media.

With the assumption that chemiluminescence contributes
to single photon events, we made the following observation
in the "singles" or no coincidence mode (Table II).

(i) Glass counting vials after dark adaptation gave a
consistent, repeatable singles count rate. Photoluminescence
can be induced by brief exposure to light due to photo-reactivation; it decays rapidly within minutes to background.

(ii) Neat solvents including 30% H_2O_2 gave only slight
or no additional singles count rate.

(iii) Tissue solubilizers such as Hyamine, Soluene,
NCS, and Protosol yielded higher singles count rates than the
solvents, and the intensity of the count rate appeared to be

Table II
Single-Events Rates of Liquid Scintillators and Accessories[a]

Sample[b]	Single-event rate[c] (cpm)	Coincidence rate (obs) (cpm)	Accidental Coincidence (calc)[d] (cpm)
Glass vial	1.09×10^4	28	
H_2O	1.15×10^4	38	
H_2O_2	1.90×10^4	–	
BBS-3	2.22×10^4	925	
BBS-3 + H_2O_2	7.17×10^4	811	1
NCS	2.2×10^5	--[e]	
NCS + H_2O_2	5.2×10^6	41	6760
Protosol + H_2O_2	3.7×10^6	55	4322
Soluene-100	5.6×10^4	20	1
Soluene-100 + H_2O_2	1.16×10^6	184	336
Triton X-100	3.7×10^6	91	3422
Benzoyl peroxide (1.5%) in toluene	1.2×10^5	44	4
Aquasol	1.71×10^5	--[e]	
Aquasol + H_2O_2	2.32×10^5	20	13
Monophase	2.0×10^5	54	10
Monophase + H_2O_2[f]	2.65×10^6	4985	1755
Monophase + N_2[g]	2.6×10^4[h]	49	
Monophase + air	7.1×10^4[h]	56	1
Monophase + O_2[g]	5.6×10^5[h]	352	78
Hypochlorite + H_2O_2[i]	1.4×10^5	24	5

a. All measurements were made on Beckman LS-9000 Spectrometer unless stated otherwise. The author is grateful to Dr. D.L. Horrocks of Beckman Instrument Co. for assistance.
b. All samples were low-potassium glass vials with aluminum cap. The sample volume was 10 ml, and 0.1 ml of 30% H_2O_2 was added. Samples were obtained commercially: Aquasol was from New England Nuclear; Monophase, Soluene-100 and Triton X-100 were from Packard Instrument Co.; NCS from Amersham/Searle; BBS-3 from Beckman Instrument Co.
c. Counted in single-photon or no coincidence mode.
d. Calculated according to the formula 2 $(N/2)^2$, where is the resolving time of the coincidence gate in minutes and N the observed single-event rate in counts per minute.
e. Not measured.
f. The content of the counting ampoule was bubbled with N_2 or O_2 gas before sealing.
h. Measurement was made on a Packard Model 3375 spectrometer.
i. The sample consisted of 10 ml of water and 0.1 ml of each reagent.

characteristic of the solubilizer. The emulsifier BBS-3 (neat) yielded a high rate on account of its high viscosity. The count rate was diminished when diluted with toluene. The decay rate of chemiluminescence in these agents appeared to be comparable. (Fig. 2).

(iv) The emulsifier-scintillator, Monophase (Packard), showed intense chemiluminescence after the addition of H_2O_2 (100/1, v/v). Accidental coincidence count rate was 3 to 4 times the value predicted by theory ($2\tau N_1 N_2$). Measured pulse height spectrum indicated the presence of multiphoton events (Fig. 3). The underlying cause is not known.

(v) Aliquots of Monophase purged with N_2, air, and O_2 and sealed in counting ampoules gave singles count rate increasing in that order indicating the ability of oxygen to generate counts.

(vi) A mixture of sodium hypochlorite and H_2O_2 (0.1 ml/ 10 ml water) yielded a high singles count rate. The mixture is known to be chemiluminescent due to the presence of singlet oxygen ($^1\Delta_g$). The singles rate observed is presumably a combination of the reaction rate and the decay of singlet oxygen.

The above observations lead us to propose that singlet oxygen ($^1\Delta_g$) is the cause of chemiluminescence in the liquid scintillation system. Oxygen is ubiquitous and singlet oxygen has a radiative lifetime of about 45 minutes [39] but has a lifetime of about 0.1 sec in the air because of quenching by nitrogen [40]. Its lifetime in solution varies with the solvent. Oxygen is soluble in toluene, and its molecular pairs ($2[^1\Delta_g]$, $[^1\Delta_g + ^1\Sigma^+]$, $2[^1\Sigma^+]$) can provide excitation energy to some fluorescent compounds in a step-ladder fashion. The implication of singlet oxygen as a source of excitation minimizes the difficulty of explaining the persistence of low-level chemiluminescence which remains above background rate over extremely long periods of time.

It may also be postulated that alkalinization of the liquid sintillation medium with tissue solubilizers induces O_2 to form superoxide ion O_2^-, which may react with a cation to form additional singlet oxygen and increase the intensity of chemiluminescence. Polymeric molecules such as surfactants, etc. may also be postulated to exert a catalytic effect in enhancing the chemiluminescence; this may account for the multi-photon events observed in the mixture of Monophase and H_2O_2. The hypothesis role of oxygen in chemiluminescence is further strengthened by our observation that the presence of stannous chloride in the mixture of Monophase and H_2O_2 enhances the intensity of chemiluminescence, presumably due to the fact that the metal ion catalyzes the decomposition

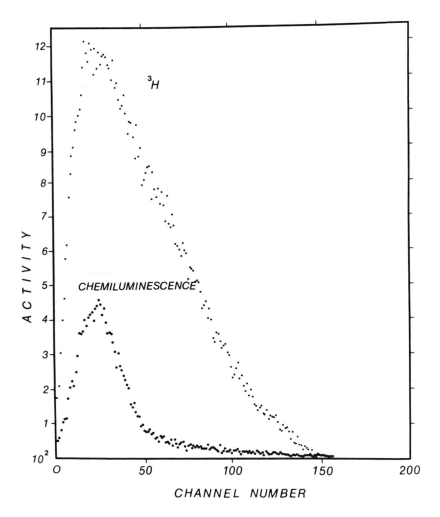

Figure 2. Pulse heights spectra of the chemiluminescent mixture (Monophase + H_2O_2) and ^3H.

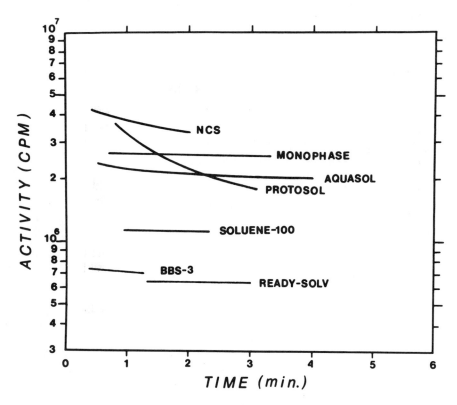

Figure 3. Decay of chemiluminescence in solution. Measurement began shortly after addition of 0.1 ml 30% H_2O_2 to the sample. The mixture (Monophase + H_2O_2) was aged for about 24 hrs.

Table III. Phosphorescence in Liquid Scintillation Components and Counting Vials.

Scintillation solvents

```
Toluene    (-)
Triton X-100 (+)
Triton X-100
  + toluene  (+)
```

Liquid scintillators

```
PPO in toluene (-)
BBOT in toluene (-)
Dioxane based
 scintillator  (+++)
PCS        (+)
Oxifluor (+)
Insta-gel (+)
Monophase (++)
```

Solubilizers

```
NCS, neat (-)
Soluene, neat (-)
Hyamine, neat (-)
```

Counting Vials

```
glass (-)
Nylon (++)
Polyethylene (++)
White plastic caps (++++)
Black caps*    (-)
```

---

*White plastic cap is spray-painted black.

of H_2O_2. This observation casts doubt on the effectiveness
of $SnCl_2$ as an agent in suppressing chemiluminescence as
previously reported [38].

Many chemiluminescent mixtures show photoluminescence;
even the mixture of ClO^- + H_2O_2 exhibits twice the singles
count rate of the glass counting vial upon excitation. As
photoluminescence occurs frequently in scintillation components and is related to the presence of luminescent impurities, it can be detected by activating the object under study
directly with an electronic photoflash. Table III shows the
phosphorescence of the scintillation components.

The singlet oxygen hypothesis provides a basis for further study on chemiluminescence in liquid scintillation systems. Photo-activation and counting of singles provides the
means for detecting and measuring minute concentrations of
luminescent impurities that are present in scintillation
solvents, additives and fluors that may diminish the
efficiency of a liquid scintillation system.

References

1. R.A. Lloyd, S.C. Ellis and K.H. Hallowes, Symposium on
    the Detection and Uses of Tritium in the Physical and
    Biological Sciences, Vienna, 1961, Vol. I. p. 263,
    IAEA, Vienna (1962).
2. R.J. Herberg, Science, 128, 199 (1958).
3. I.B. Berlman, Handbook Fluorescence Spectra of Aromatic
    Molecules. Academic Press, New York (1965); 2nd Edition,
    (1971).
4. J.B. Birks, Photophysics of Aromatic Molecules, Wiley-
    Interscience, New York (1970).
5. M.W. Windsor in Physics and Chemistry of the Organic
    Solid State (D. Fox, M.M. Labes and A. Weissberger, eds.)
    p. 343 Wiley-Interscience, New York, (1965).
6. F. Wilkinson, Quart. Rev., 20, 403 (1966).
7. D.C. Horrocks, Nucl. Instrum. Methods, 128, 573 (1975).
8. M. Inokuli & F. Hirayama, J. Chem. Phys., 43, 1978
    (1965).
9. H. Ishikawa and M. Takiue, Nucl. Instrum. Methods, 112,
    431 (1973).
10. K.D. Gundermann, Angew. Chem., Internat. Edit. 4, (7)
    566 (1965).
11. F. McCapra, Quart. Rev., 20, 485 (1966).
12. M.M. Rauhut, Acc. Chem. Res., 2, 80 (1969).
13. E.H. White and D.F. Roswell, Acc. Chem. Res., 3, 54
    (1970).

14. S.R. Abbott, S. Ness and D.M. Hercules, J. Am. Chem. Soc., 92, 1128 (1970).
15. D.M. Hercules in The Current Status of Liquid Scintillation Counting (E.D. Bransome Jr., ed.), p. 315, Grune & Stratton, New York (1970).
16. A.U. Khan and M. Kasha, J. Amer. Chem. Soc., 92, 3292 (1970).
17. L.J. Bollyky, R.H. Whitman, B.G. Roberts and M.M. Rauhut, J. Amer. Chem. Soc., 89, 6523 (1967).
18. H.F. Cordes, H.P. Richter and C.A. Heller, J. Amer. Chem. Soc., 91, 7209 (1969).
19. M.M. Rauhut, D. Sheehan, R.A. Clarke and A.M. Semsel, Photochem. Photobiol., 4, 1097 (1965).
20. D.L. Maricle and A. Maurer, J. Amer. Chem. Soc., 89, 188 (1967).
21. A. Zweig, D.L. Maricle, J.S. Brinen and A.H. Maurer, J. Amer. Chem. Soc., 89, 473 (1967).
22. C.S. Foote, S. Wexler, W. Ando and R. Higgins, J. Am. Chem. Soc., 90, 975 (1968).
23. R.W. Murray and M.L. Kaplan, J. Amer. Chem. Soc., 90, 4161 (1968).
24. H.H. Wasserman and J.R. Scheffer, J. Amer. Chem. Soc., 89, 3073 (1967).
25. E. Rapaport, M.W. Cass and E.H. White, J. Amer. Chem. Soc., 94, 3153 (1972).
26. M. Kasha and A.U. Khan, Ann. N.Y. Acad. Sci., 171, 5 (1970).
27. E.J. Corey and W.C. Taylor, J. Amer. Chem. Soc., 86, 3881 (1964).
28. R.M. Arneson, Arch. Biochem. Biophys., 136, 352 (1970).
29. R.M. Howes and R.H. Steele, Res. Commun. Chem. Pathol. Pharmacol. 3, 349 (1972).
30. R.C. Allen, R.L. Stjernholm and R.H. Steele, Biochem. Biophys. Res. Commun., 47, 679 (1972).
31. K. Goda, J. Chu, T. Kimura and A.P. Schaap, Biochem. Biophys. Res. Commun., 52, 1300 (1973).
32. D.A. Kalbhen, Int. J. Appl. Radiat. Isotop., 18, 655 (1967).
33. A.A. Moghissi in The Current Status of Liquid Scintillation Counting (E.D. Bransome Jr., ed.), p. 86, Grune & Stratton, New York (1970).
34. D.A. Kalbhen and A. Rezvani, Organic Scintillators and Liquid Scintillation Counting, Proc. Int. Conf. 1970 (D.L. Horrocks, C.T. Peng, eds.), p. 149, Academic Press, New York (1971).

35. D.A. Kalbhen, Liquid Scintillation Counting, Vol. 1
    Proc. Symp. 1970 (A. Dyer, ed.), p. 1 Heyden, London
    (1971).
36. D.S. Kearns, Int. J. Appl. Radiat. Isotop., $\underline{23}$, 73
    (1972).
37. R.D. Neame, Anal. Biochem., $\underline{64}$, 521 (1975).
38. E.D. Bransome, Jr. and M.F. Grower $\underline{in}$ The Current
    Status of Liquid Scintillation Counting (E.D. Bransome,
    Jr., ed.), p. 342, Grune & Stratton, New York (1970).
39. B. Stevens, Acc. Chem. Res., $\underline{6}$, 90 (1973).
40. B. Stevens, Personal Communication.

Discussion

Dr. B.E. Gordon - Lawrence Radiation Laboratory (US)

I have some concern about two of the introductory talks, one by Dr. Wigfield and one by Mr. Ferris. I am actively concerned about the application of solubilizers to biological samples and also about the possibility that we have unknown systems, by which I mean that the composition and the exact structure is unknown.

I will talk about the adsorption problem first and then about solubilization.

It is true that you can do some tests to determine whether you have an adsorption problem. It is incorrect, if you have an unknown system and you observe adsorption, to apply a carrier addition approach because this approach will only work when you know the structure of the adsorbed molecule. I must also express my grave concern about the adsorption shift as a quantitative method. That will work I think again only when you know the exact structure of the molecule.

Now as to the problem of solubilization, I think a brief lecture in polymer physics is in order. It is well known that non-polar polymers such as polybutene irreversibly adsorb onto active surfaces when they exceed a certain molecular weight. This is because the possibility that all adsorbed points of the randomly coiled polymer have a low probability of being desorbed at the same time. As the molecular weight increases the probability decreases; all this occurs even if the heat of adsorption per monomer unit is small. It is clear then that the situation with polar molecules (i.e. proteins, DNA, tissues, etc.) is much worse. The molecule can be much smaller and still be irreversibly adsorbed because the heat of adsorption is much greater per unit monomer (e.g. amino acid). Imagine then a macromolecule of 1000 monomer units which has only 10% of its polar sites available for adsorption and the heat of adsorption is 2 k cal/mol. The molecule is adsorbed in a dynamic state with some groups desorbing and others adsorbing but the probability of all adsorbed groups coming off at once is exceedingly small. As long as one group is adsorbed, then the molecule is held close to the vial wall and 4π geometry is not achieved. Looked at another way, the energy required to ensure simultaneous desorption of all 100 groups from the surface is 2 x 100 or 200 k cal/mol - greater than C-C bond strengths and so the molecule is irreversibly adsorbed.

When you apply the solubilizer, whether it be an acid or a base or a quaternary ammonium compound to that kind of a system and it is applied to a macromolecule, the weakness of the situation is that one has no idea of what the species are like. If one has no idea, then one cannot predict whether the measured count rates will be accurate. If any part of the system is adsorbed, and if the molecular weight is high enough, then by the explanation above some of this is irreversibly adsorbed and a correction cannot be made.

I would like to support Dr. Bransome wholeheartedly in pointing out that if the material you have to deal with is water soluble then you can count it in a detergent system. If it is fat or oil soluble it can be counted, and counted accurately I might say, in a toluene based fluor. All other materials as far as I am concerned must be burned. I have refereed some papers on this and I must say that if I am called on again, and if the sample preparation is by way of solubilization, I will have to turn in a negative report.

The advantage of combustion is not so much that the sample is converted into a form that is easily handled but that you know the molecular composition of the labelled species. When you know that, whether it be a hydrocarbon or a carboxylic acid or whatever, then you can deal with it. On the other hand, when the molecular composition is unknown our measurements are really fraught with danger.

The only advantage I can see in solubilization is that you handle the samples in parallel rather than in series as you do with combustion. Everything else I hear about it is in fact negative.

Dr. D.C. Wigfield - University of Victoria (CAN)

I agree with both points that Dr. Gordon made about adsorption problems. First of all, clearly if you do not know what the material is that you are counting, and if it does not have a well defined structure then this is one of the biggest limitations of the carrier dilution technique. Secondly, I also agree that the adsorption shift technique may be questionable when you do not know the chemical structure of the thing you are trying to count.

Dr. A.A. Noujaim - University of Alberta (CAN)

Dr. Wigfield, I have read your elegant articles concerning adsorption. However, the double ratio technique you describe applies only to ^{14}carbon. If I recall correctly, one of the compounds that is severely adsorbed is glucose. Now glucose is used extensively in biochemistry, especially in the tritiated form. Is there a way by which we can predict adsorption of glucose or other tritiated compounds to glass?

Dr. D.C. Wigfield

If you are asking the question, is there a way to overcome adsorption for tritiated materials, I do not know of any except the classical way of getting rid of it by carrier addition or something like that.

If you are asking about the question of detection of adsorption which I think was your question, then I would think that carrier dilution and vial emptying might give you some kind of hint. We have not applied adsorption shift to tritiated materials.

Dr. E.D. Bransome Jr. - Medical College of Georgia (US)

When Dr. Wigfield published his first article on this topic we immediately went to try it with tritium, and as a matter of fact one of the things we used was tritiated glucose. It simply does not work. One loses the counts altogether.

Dr. K. Painter - Colorado State University (US)

I would like to comment on an additional couple of ways reported in our review on counting vials in the Sydney symposium (1973). When using glass vials siliconizing the walls helps. It is a messy procedure but almost always works for everything. Another possibility is pH change. The detergents are usually quite good at preventing things like glucose from adsorbing. Lastly, complexing agents, and there are many of these, 2-ethylhexanoate, the quaternary ammonium compounds etc. which are all used. There are about ten ways actually.

Dr. A.A. Noujaim

I address this question to Dr. Bransome. I would like to object to the use of the term DPM in publications because it relates to absolute activity while what we use when we measure samples is a secondary standard, ±5%. I am wondering whether we could use the term relative DPM, or relative CPM, or standardized CPM rather than absolute DPM as is implied at present.

Secondly, in our experience the use of different standards from different manufacturers unfortunately gave

different results. Is there any particular experiment that
we should use to standardize those supposed standards?

Thirdly, I would like to hear some comment about the
present status of counting ^{125}iodine in liquid scintillation
counters.

Finally, how do we check current commercial solubilizers?
Should we use some simple criterion such as the double ratio
method to indicate the maximum permissible water load rather
than the figure of merit which could be very deceptive?

Dr. E.D. Bransome, Jr.

Regarding your comments about DPM, there are no acceptable absolute standards. It seems to me that for some
isotopes the National Bureau of Standards in the United
States might be encouraged to expand its efforts and
activities to some degree. One might therefore be able to
relate experimental findings to one of those absolute or
'agreed upon' standards.

I would submit that provided we define what we mean by
DPM that too would be probably acceptable. I think your
implied criticism is correct and that few of us do make such
a definition.

Regarding solubilizers, it is quite obvious from your
work, and some of our own observations that back you up
though not as elegantly, that the double ratio plot is not
a significantly rigorous procedure. It is certainly not
rigorous enough to make sure that a sample loaded with
surfactant is homogenous.

Dr. Bush-Mueller's suggestion that the same isotope be
put in tracer amounts in both phases, that is organic solvent
and water, seems in some cases to provide an additional
criterion with a little more power. The truth of the matter
is that we do not know enough, and not enough work has been
done in this area to really answer the question. I am sure
you know that and you are asking the question to ensure that
there is continuing attention to the problem.

As far as ^{125}iodine counting is concerned, one can count
the conversion electrons, but one must be aware that the
usual quenching effects on counting efficiency will be somewhat more dramatic than that observed for tritium which has
a similar pulse height spectrum. It is a reasonable technique but it must be carried out with even more care than
counting the usual beta emitters.

Dr. A.A. Noujaim

Dr. Noakes, I must agree totally that combustion is the
method of choice and those people who have been solubilizing

tissue samples for years know the problems that are involved
in preparing tissue samples for liquid scintillation
counting. However, we have one problem which is not
completely solved. We lack an absolute biological standard
in order to compute the true recovery of the combustion
system. What is the present status regarding the availability of such a standard right now?

Dr. J.E. Noakes - University of Georgia (US)
    I understand that the instrument companies have prepared
some acetate type paper which is labelled with [14]carbon and
tritium of known activity. These small pieces of paper are
put into the combustion apparatus and burned so that recovery
may be measured. This, however, does not account for the
differences between the acetate paper and biological
materials.

Mr. E. Polic - Packard Instrument (US)
    We have developed a technique for our oxidizers. We make
a comparative check with a solution that we have developed.
You can combust it and also prepare a sample without
combustion and make a comparison between the two to check
the recovery. This procedure is described in the operation
manuals of our oxidizers.
    We encountered the same problem of finding a suitable
standard as implied in Dr. Noujaim's question, and that is
why we went to a comparative method.

Dr. K. Painter
    I'd like to suggest that before we get everyone converted
to the combustion technique that in a clinical lab it is
clearly not the answer until more work is done.
    I list below a few disadvantages of the combustion
technique when you need to process large numbers of samples.
    i)   cost per sample
   ii)   processing time per sample
  iii)   carry-over or memory
   iv)   capital expense
    v)   other nuclides.
Thus, we should not leave this conference with the idea that
combustion is a panacea. I do not think that solubilization
is a method that is free from faults either, but until such
time that there is a method to do large numbers of samples
quickly and inexpensively, I do not think that combustion
will catch on.

Dr. E.D. Bransome Jr.

Since I am one of the advocates of combustion I should like to remark that I am not so sure that Dr. Painter's objections to combustion in the clinical labs are really terribly cogent right now because there are not many clinical procedures using liquid scintillation counting, although the future is probably going to offer many more. I will ask what about other isotopes, some of which have already been mentioned at this meeting; Plutonium isotopes present in bone, ^{35}Sulphur and a variety of labels that have not been proven to carry over, or to be recovered well in the presently available combustion systems. I do know that there are some ways to achieve combustion of samples that are not tritiated or containing ^{14}carbon and make this comment to hopefully elicit some information on the subject.

Dr. J.E. Noakes

It is unfortunate that Dr. Kisieleski is not here because he has done quite a bit of work with some of the other isotopes in combustion. Talking with him earlier, he said he has worked with ^{35}sulphur and ^{125}iodine and that the recoveries were not at all quantitative, about 80% in one case.

I would say that for isotopes other than tritium and ^{14}carbon that there is a big question which will have to be worked on. For ^{59}iron I might point out that the Searle 6550 combustor, which uses a planchet to hold the sample, collects any residue which is not combustible on that planchet. This might be a way of collecting a third isotope in the sample since it would not be carried over into the trapping mechanism. This will have to be looked at as I do not think there is any data available on this at present.

Regarding sample throughput, the combustor is well automated and takes between one minute and one minute and a half to process a sample which is quite fast. But it does require an operator even in an automated stage to tend the samples and place them in the combustor and cap the vials afterward. Looking at liquid scintillation counters with say a three-hundred sample capacity, reveals that it is the combustor that limits throughput. On the other hand it is possible we might see another generation of combustors to come out in which we would combust the samples, transfer directly the combustion products into a flow cell to be counted, and then discharge the waste into a receptacle. In this way we could do combustion and counting in a very rapid mode. Maybe this will occur, but right now we are at the single sample combustion stage. If we compare this to the Schoniger flask which took ten to twenty minutes per sample, then the combustors of today are a great improvement, and certainly expedite the preparation of samples.

Dr. S. Apelgot - Institut du Radium (F)

The only trouble with the combustion technique is memory. Ordinarily one does not know the activity of experimental samples and if you happen to burn one of low activity after one of high activity you can have some difficulty.

One difficulty with the application of the solubilization technique to tritiated compounds in vivo is that one of the metabolic products is water. When you neutralize the samples after solubilization there is an increase in temperature which may vaporize that water.

Dr. J.E. Noakes

The memory of the available combustion units is low because there is a steam purge through the system to drive any residual material over into the collecting vial. Then after the delivery of the cocktail is completed, there is a second steam purge which cleans the whole system out. Typical recovery for tritium might be 99%, and this is valid for real samples as well as filter paper. I would say that the manufacturers have done a pretty good job in eliminating the memory. There is some spillover when doing dual labelled samples but this is very much less than 1%.

Dr. S. Apelgot

In my work with glass fibre filters I needed an accurate technique to check my own methods and I therefore used combustion. We came to the conclusion that it was necessary to burn an additional glass fibre filter between samples to be sure there will be no contamination from one sample to the next. Even 1% from a very high activity may give you enough contamination to cause large errors if the activity of the next sample is very low.

Dr. B.E. Gordon

The main concern for memory in the Packard 306 is in the tritium channel because the memory in the ^{14}carbon channel is extraordinarily low. If one burns a tritium sample containing 100,000 DPM then the memory after the

purging process that Dr. Noakes has described is only
five counts per minute above background. We regard that
as an acceptably low value. If anybody is concerned about
those five counts per minute when I would suspect that the
samples would have to be pre-screened.

I would like to ask a question of Dr. Noujaim. I do
not understand the question about wanting to burn a biological standard in the oxidizer as compared to any standard.
If a sample is burned to carbon dioxide and water, it eludes
me as to why you are concerned about the original composition
so long as it is burned in a quantitative way. We burn
both biological samples and hydrocarbons and have not had any
problems. I wonder if you would respond to that.

Dr. A.A. Noujaim

My question relates to my concern about the completeness
of the combustion. We have no control over the temperature
which may change from time to time. If there is an incom-
plete combustion we may not get a complete recovery. We
may not obtain carbon dioxide for example.

Dr. J.E. Noakes

If you look at the combustors they are very versatile.
You can regulate the oxygen flow, you can regulate in some
cases the temperature of the reaction, you can use extra fuel
material to slow the reaction down, you can determine how
long you want it to combust, there are a lot of variables
that you can adjust. If you combust a sample that burns very
fast and it does not get enough oxygen then you will get a
dirty burn and you will have problems. We are not saying that
the combustion units that are on the market today, and I
think I speak for all of them, are completely automated so
that you can throw anything in and expect to get a clean
burn every time. I think you have to apply a rational
approach just as in any analytical technique.

If you have a lot of samples that are very similar
then once you have set the conditions up appropriately then
you can expect to get consistent results. You can adjust
all the flow rates, what size sample you want, what type of
cocktail, the combination of cocktails using unlabelled
material.

If you change the nature of the samples then you must
go back and find new optimal operating conditions. There
is some operator expertise that goes with it too.

Dr. D. Horrocks - Beckman Instruments (US)

I must defend myself from some of Dr. Bransome's earlier comments and to talk a little bit about two things. First of all about the 'H' number as compared to the limitations of External standard channels ratio (ESCR) and sample channels ratio (SCR) methods. Secondly, I will provide a little information on [125]iodine counting in liquid scintillation systems.

Concerning this latter topic, just this last month a paper of mine has been published in Nuclear Instruments And Methods (volume 133 page 293 - 1976) on "The measurements of [125]iodine by liquid scintillation methods". I have analyzed the mode of decay of [125]iodine as it relates to the emission of auger electrons and conversion electrons which would be detected in a liquid scintillation system. Also I tried to get some estimate of the contribution of the x-rays and γ-rays to the counting efficiency of [125]iodine in a typical liquid scintillator. In this paper we have done some studies on the counting efficiency with different types of quenching, and also have made a measurement of the pulse height spectra.

The spectrum of [125]iodine exhibits two peaks spread over a range of pulse heights somewhat wider than that usually obtained for tritium. So the [125]iodine will have an appreciable counting efficiency even for such a degree of quenching that would have reduced the tritium counting efficiency to zero. This is because for [125]iodine we have a fair number of electrons which have energies greater than the end point energy of tritium. We also found out that in a twelve milli-litre volume of liquid scintillator less than eight percent of the x-rays or γ-rays interacted with the scintillation cocktail at all. They comprised a very small contribution to the counting efficiency.

Dr. E.D. Bransome Jr.

In our work, some of which is represented in an Analytical Biochemistry Article published in 1973, when we were looking at high levels of quenching we were no longer able to resolve the two peaks of [125]iodine. I have no argument with what you have to say about being able to completely quench tritium and still get some [125]iodine counts. However, the quench correction curve was rather different in our hands than for tritium and with high levels of quenching we found a rather more precipitous fall and this is what I was referring to earlier. This was probably due to the loss of the peak at lower pulse heights, but we couldn't tell which because of lack of resolution.

Dr. D. Horrocks
In the article I referred to there is a figure showing
the effect of both colour and chemical quenching on the
pulse height spectra of ^{125}iodine in a liquid scintillation
system. The effect of quenching is rather similar to
increasing the lower threshold and indeed over the initial
levels of quenching the ^{125}iodine count rate will decrease
quite dramatically because you are eliminating a fair
number of the low energy electron events. At higher levels
of quenching you will have a more gradual decrease. We
constructed some quench curves using something similar to
an 'H' number and they were practically linear. Perhaps
in the ESCR or the SCR methods you may very well get some
funny quench correction curves but by using this method it
straightens things out and gives a very nice line.
Now I would like to clear up some uncertainty with regard
to the limitations of methods of quench correction which are
based on a measurement of pulse height and I think Mr. Laney
will agree with this since Searle uses a similar technique.
When you use these techniques you are not as limited in the
capability of being able to make measurements over a much
wider range of quenching as you would be using the ESCR or
SCR methods. Also, if you use the ESCR or SCR methods you
can get (almost) any shape of curve you want depending upon
how you choose your windows. Therefore, it is not difficult
to understand how people in different parts of the world have
a hard time comparing results when you consider the variability in ESCR or SCR calibration curves. Now if you are
measuring a pulse height, or an average pulse height, you
are measuring something which is invariant. Thus, quench
correction curves will be reproducible from lab to lab.
The ESCR method requires two counting windows to obtain
the necessary ratio values and these windows may be selected
in many ways. Typically you might choose two overlapping
windows, one (window A) covering a wide range of pulse
heights and a second (window B) that covers a narrower range
of pulse heights because its lower level discriminator is set
at a higher value. The required ratio may then be computed
as the ratio of the counts in those two windows due to some
external standard source. Now at high levels of quenching
the range of pulse heights due to the external standard may
all fall below the lower level of window B giving a ratio
(B over A) of zero. At even higher levels of quenching you
will also get a ratio of zero and you can no longer inter-
polate the sample efficiency from the quench correction

curve. Now if you measure a pulse height value you do not have that limitation. Limitations only arise when the compton spectrum is completely removed which is a quench factor of some four-hundred, and much greater than the twenty or thirty that we can obtain with the ESCR or SCR methods.

Also, recently we examined a problem in using the ESCR method when counting in plastic vials. We found that as the plastic vial aged with the liquid scintillator in the vial, the scintillation cocktail and the sample and the solutes would diffuse into the plastic wall. In doing so, it converted the plastic wall into a plastic scintillator because it deposited the scintillators into the matrix. When the external standard was brought up next to the vial compton electrons were created in both the scintillation cocktail and in the plastic vial wall. Since the additional scintillators due to the plastic wall are detected as a distribution of counts at small pulse heights, the spectrum of the external standard becomes distorted by an increase in counts at lower pulse heights.

Thus, you might get additional counts in channel A but not in B, giving an external standard ratio which changes with time as the diffusion takes place. Depending on how the windows have been selected the ratio may start to change either as soon as the sample is prepared or it may start to change up to three days later. That is, the higher the lower level of window A, the longer it will take for the diffusion to cause a problem. But the more you raise the lower level of window A, so the lower level of window B must be raised a corresponding amount and the range of quenching that can be handled by the ESCR method is thereby reduced.

One point I want to make is that none of the additional counts I have mentioned occur in the region of the compton edge. Thus, the 'H' number is not influenced by this effect.

Dr. K. Painter

I guess one could summarize Dr. Horrock's comments by saying that the ESCR method that has been thrust upon us by commercial people for ten years has now fallen into disrepute.

Dr. J.A.B. Gibson - AERE Harwell (UK)

I would like to support what Dr. Horrocks has to say about the 'H' factor being independent of energy. The following table shows that the relative quenching factor

which I call 'G' is the same for a wide range of isotopes at
the same quenching level. In fact this tends to support all
that Dr. Horrocks has said.

Relative Quenching Factor For A Wide Range Of Electron And
Beta Ray Energies

Isotope	Type	Energy Beta Max KeV	Relative Quenching Factor G
^3H	beta spectrum	18.5	0.80
^{14}C	beta spectrum	115.0	0.82
^{137}Cs	compton edge	478.0	0.82
^{137}Cs	conversion line	625.0	0.80
^{36}Cl	beta spectrum	707.0	0.81
^{207}Bi	conversion line	980.0	0.85

Mean relative quenching factor = 0.82 ± 0.02
(from J.A.B. Gibson. Int. J. Appl. Rad. Isotopes 18, 681,
1967).

Dr. E.D. Bransome Jr.
I would like to respond to one of Dr. Horrock's first
comments. When I talked about the potential limitations
of the new methods of quench correction I did in no way
refer to the dynamic range, but to the validity of these
methods as methods of standardization. I would like to re-
emphasize that I think that the limitations, as with the
older methods of ESCR and SCR, are going to be related to
sample homogeneity, and this is what I was referring to.
I think for the group present here this is a very obvious
matter but for the majority of instrument buyers it will be
an additional problem.

Mr. E. Polic
Some of Dr. Horrock's comments on dynamic range depend
on the radionuclide used for the external standard. You can
get around some of these problems by using a higher energy
radionuclide for the external standard.
In Dr. Painter's introductory remarks he had a table
which showed figures of merit for various cocktails. I
think one should be cautious about comparing those figures
of merit because they were taken over several years and
because there has been a lot of instrumentation changes
that occurred over that period of time. If you really

want to make a comparison of those cocktails you should measure them all in a single experiment using a single instrument, instead of taking data from a ten year period.

I liked the bibliography you sent around during your introductory talk but it looks like there is a void between 1957 and 1969, and there were several liquid scintillation symposia during that period. There was one in 1960 (Albuquerque),[1] there were a number of symposia that were organized by Packard and New England Nuclear, and Plenum Press published four volumes on the subject. I think they were called 'advances in tracer methodology' and are still available.[2] New England Nuclear had a number of good application articles in their Atom Light which is no longer available.

---

[1] Daub, Guido H.; Hayes, F. Newton; and Sullivan, Elizabeth, Proceedings of the University of New Mexico Conference on Organic Scintillation Detectors, U.S. Government Printing Office, TlD-7612 (1960). (1960 Symposium, Albuquerque).

[2] Rothchild, Seymour, Advances in Tracer Methodology, Volume 1 (1963), Volume 2 (1965), Volume 3 (1966), Volume 4 (1968), Plenum Press.

Mr. T. Horan - W.W. Cross Institute (CAN)
I wish to address my question to Dr. Peng, concerning the singlet oxygen generating system of hydrogen peroxide/sodium hypochlorite. Upon the mixing of these two chemicals there is a red flash due to the production and electronic relaxation of singlet oxygen. This red flash has been measured at 634 nanometres and lasts for one tenth of a second. We have measured a much longer lived emission from this system. However, the wavelengths do not match the emission pattern expected for singlet oxygen. I ask how you can be absolutely certain that this chemiluminescence is that produced by singlet oxygen. If it is singlet oxygen, why is there a shift in the observed wavelengths from the expected wavelengths?

Dr. C.T. Peng - University of California (US)
I can answer your question by giving you some of the absorption spectrum of, for instance, the hydrated electron. You have a solvent effect where the absorption maximum changes due to this effect. In the case of singlet oxygen

it has been reported in the chemical literature that
hypochlorite with hydrogen peroxide would give singlet
oxygen. I think this has been studied extensively by a
number of schools, (especially Kasha and Khan) who have
given some of the absorption spectra of these molecular
species. Of course the light that you observe at 600
nanometres is probably associated with a very rapid decay.
We have used the new Beckman spectrometer with which you
can measure the decay of the single photon, but this is a
very complicated reaction because what we are observing
is the rate of the reaction between the two reactants and
also the decay of the singlet oxygen. The singlet oxygen
itself has a radiative lifetime of forty-five minutes.
But in air the proportion of singlet oxygen is not known
and it has a lifetime of approximately one tenth of a
second, because of the quenching by nitrogen molecules.
The singlet oxygen has a variable lifetime in different
solvents. For instance, in carbon disulphide, I think
Kearns has measured the lifetime in to be about seven-hundred
microseconds. So far as your observation is concerned, I
have not come across anybody measuring this directly, that
is for singlet oxygen using this oxidation mixture. I should
mention that singlet oxygen is not only generated by the
reaction of hypochlorite and hydrogen peroxide, it has been
observed in a number of reactions. Also, microwave discharge
through gaseous oxygen and some of the enzyme systems also
give rise to singlet oxygen.
    I think Dr. Schram mentioned FMN and NADH systems where
oxygen is involved. Probably those peaks, the λ_{max}, are due
to singlet oxygen. I would say that what you observed
probably is due to a slow component of the singlet oxygen.

Dr. W. Reid - University of Saskatchewan Hospital (CAN)
    Do the commercial liquid scintillation counters provide
a suitable means of comparing the observed accidental
coincidence rate to that which is calculated from the singles
rate? That is, can we really eliminate the pulse height
analysis part of the machine? Can you test the machine with
truly incoherent light? I think perhaps one can by perhaps
introducing a light leak or a small flashlight.

Dr. C.T. Peng
    All the information that I presented in that table
earlier indicates the current state of the art. As I

understand it, the new Beckman machine is able to give you
the singles rate and also at the same time give you the
accidental coincidence rate. We have made some measurements
with a Packard machine by switching between the singles and
coincident modes and noticing the difference. As far as I
know, Dr. Ross has shown that fluorescence of a glass vial
which has been excited with coherent light is good as a
means of monitoring the scintillation purity of components.

Dr. S. Apelgot
     I would like to return to the topic of glass fibre
supports. I have worked for many years with this technique
under different experimental conditions and I feel it is a
good technique because it is reproducible, and samples are
stable. It is true that I have to construct special
calibration curves, but whenever I try homogenous solutions
I have greater difficulties and I return to fibre glass
supports. Even with tritium, and even if the labelled
compounds are not dissolved by the liquid scintillator, I
believe that the fibre glass supports provide a superior
technique.

Dr. E.D. Bransome
     In reply to Dr. Apelgot's comment all I can say is that
in our hands we encountered sufficient difficulties with a
variety of samples, even those prepared on non-absorptive
supports to render the technique undesirable from our point
of view.

THE CLOSING ADDRESS OF THE CONFERENCE GIVEN BY
Dr. P.E. Stanley

At the last conference which was held in Sydney,
Australia nearly three years ago, Dr. J. Birks had the prive-
ledge of summing up at the end of that conference and there
is no way that anybody could outdo those comments he made at
that time.   I can recommend for those of you who haven't
read his concluding remarks that they are well worth read-
ing and are in the last few pages of the proceedings of that
meeting (Liquid Scintillation Counting: Recent Developments,
Academic Press, 1974).
     Now let's get to this conference.   The high scientific
standard which has been set at the previous meetings has
not only been maintained but some new ideas and concepts
have been injected.   Firstly, I think most people would
agree that the discussion has been much more vigorous than
we have seen at any other conference, at least the ones I've
been to, and I've been to most of them over the past five or
six years.   Secondly, a good deal of the talk has centred
around our responsibilities in keeping the everyday user of
liquid scintillation counting informed about the present
state of the art, not only in regard to the equipment but
also sample preparation and its inherent problems.
     Another important item is that a steering committee,
tentatively titled this morning as the 'International
Organisation for Liquid Scintillation Science and Technology'
has been set up.   The committee consists of: Dr. S. Apelgot,
Dr. K. Painter, Dr. B.E. Gordon, Dr. E.D. Bransome, Dr. A.A.
Noujaim, Dr. J.A.B. Gibson, Dr. E. Schram, Dr. J. Noakes
and Dr. P.E. Stanley(chairman).   We will try to look after
the long term interests of the liquid scintillation counting
fraternity, the forward planning of future conferences,
and possibly in the future, the setting up of a formal
society to bring together interested workers from the diverse disciplines in which liquid scintillation counting is
used.
     As is usual in these conferences, we've listened to a
very wide range of papers.   Some topics included the scintillation process itself, Cerenkov counting, the counting of
novel radionuclides, quench correction etc.   Our thanks of
course are extended to the plenary lecturers: Dr. D.L.
Horrocks, Dr. J.A.B. Gibson, and Dr. H.H. Ross and perhaps
even myself!   We are very sorry that Henry Polach was unable to be with us due to illness just prior to his depar-
ture from Australia.

The workshops have proved, I think, to be really excellent. They've provided a forum for all sorts of discussion between the delegates. It is my opinion that at future meetings more time should be allocated to the workshop sessions. We've spent about four hours this time and I think in the future we should perhaps add a couple of more hours to this form of discussion which I believe is so valuable for a science which is used in so many disciplines and by so many workers.

Our thanks go of course to the delegates without whom such a stimulating meeting couldn't have occurred, and to the Government of Alberta and to the University of Alberta and the Commercial Sponsors who generously provided the financial backing which makes such a conference viable in the first place. Also, we must thank the staff of the Banff Centre for making our stay a pleasant experience.

Finally, to the scientific advisory committee our thanks go for their help, but of course most of all our thanks go to Tony Noujaim, Len Wiebe, and Chris Ediss for their very hard work in organising such an excellent conference.

To all my old friends, and to my new ones, I look forward to meeting you again whenever and wherever it may be and hope that we can continue to shed some light on this technique which seems to have so many pitfalls for the unwary.

Index

A 6
B 7
C 8
D 9
E 0
F 1
G 2
H 3
I 4
J 5